The Open University

Science: A Third Level Course

Ecology   **Block B   Population dynamics**

Introduction to Block B

Units 6 and 7
Insect populations

Unit 8
Populations of plants
and vertebrates

Unit 9
General principles of
population regulation

Prepared by an Open University Course Team

THE OPEN UNIVERSITY PRESS

S323 Course Team

Chairman and General Editor

M. E. Varley

Authors

Mary K. Bell (*Staff Tutor*)
E. A. Bowers (*Staff Tutor*)
N. R. Chalmers
Irene Ridge
C. Turner (*Staff Tutor*)
M. E. Varley

Editor

Eve Braley-Smith

Other Members

N. Cleminson (*BBC*)
S. H. Cousins
R. D. Harding (*Course Assistant*)
S. W. Hurry
A. R. Kaye (*IET*)
E. Milner (*BBC*)
R. M. Morris (*Technology*)
S. P. R. Rose
J. Stevenson (*BBC*)

Consultants

J. M. Anderson
M. Burgis
K. Southern

The Open University Press,
Walton Hall, Milton Keynes.

First published 1974.
Copyright © 1974 The Open University.

Designed by the Media Development Group of the Open University.

Printed in Great Britain by
Martin Cadbury, a specialized division of Santype International,
Worcester and London.

ISBN 0 335 04152 3

This text forms part of an Open University course. The complete list of units in the course appears at the end of this text.

For general availability of supporting material referred to in this text, please write to the Director of Marketing, The Open University, P.O. Box 81, Walton Hall, Milton Keynes, MK7 6AT.

Further information on Open University courses may be obtained from the Admissions Office, The Open University, P.O. Box 48, Walton Hall, Milton Keynes, MK7 6AB.

1.1

Introduction to Block B

Contents

Introduction

Studying energy flow through ecosystems usually involves estimating biomass of the various trophic levels and to calculate this, especially for animals, it is usually essential to have information about the numbers of individuals of various species and the way in which these numbers vary from time to time. In comparing ecosystems, it is necessary to know how the numbers vary from place to place. The study of numbers within populations and communities is the province of population dynamics, and this is the aspect of ecology presented in this Block of Units.

There are three basic aims of Block B:

1   To consider the sizes of populations of various organisms and how they vary; in particular, to investigate why some species are common and others are very rare, and why numbers in some populations fluctuate, sometimes more than a thousandfold.

2   To develop rigorous criteria for evaluating the part played by various factors of the environment (physicochemical and biotic) in regulating population numbers and determining population changes.

3   To introduce some mathematical expressions and theories that describe or attempt to explain changes and stability of population numbers.

The organisms which are studied first are insects, and from these we move to other organisms—flowering plants, fishes, birds and mammals. The study of human population dynamics is deferred until Block D.

Why start with insect population dynamics?

It is pleasant to sit in an armchair and develop hypotheses, but these are of little use unless they can be tested. Hypotheses about population numbers and why they change can be tested only if sufficient numerical information is available, and this should cover as many generations as possible. Organisms with short lives are therefore better subjects for study than those, such as elephants and men, with long lives. It is convenient to be able to see the organisms with the naked eye rather than to rely on culture methods for counting and identifying them, as with bacteria. It is convenient if the organisms can be maintained in the laboratory and if they are present in large numbers in accessible habitats. Insects possess many features that make them ideal subjects for the study of population dynamics: they are small enough to handle easily in large numbers but large enough to search out in the field, and many are hardy enough for laboratory investigations. Some insects live for many years, such as the American cicadas, *Magicicada septemdecim*, which take seventeen years to grow from egg to adult; others, such as fruit-flies, *Drosophila* spp., have a generation time of two or three weeks. In temperate latitudes, many insects have a single generation each year; the adults lay their eggs and die and the other stages of the life history— the larval instars (stages separated by moults) and pupae—are spaced out through the year, eventually giving rise to the next generation of adults. Insects with an annual life cycle and generations that scarcely overlap present the investigator with certain advantages when it comes to devising methods of sampling and the numerical treatment of observations.

Another aspect of insects which makes them appropriate subjects for study is that many are of considerable economic importance, mostly as pests but some as beneficial insects. Mosquitoes (which transmit malaria), fleas, lice and other insects of medical importance have been studied from the point of view of eradication, as have such insect pests as cabbage rootfly, spruce budworm and citrus scale insects. As you will read in the prescribed book *Insect Population Ecology* (see p. 56), many of the advances in thinking about insect populations have come from studies of pests of stored products and of forest trees. These pests are of economic importance; theories about them can be tested practically and may have considerable financial implications. Government support is often available for this type of research, and manufacturers of insecticides are also greatly interested in why there are insect pests and how they should be 'controlled'.

Fish populations have been much studied because they also are of economic importance; the theory of managing fisheries and adjusting 'crop' to 'stock' has been discussed in many publications. Most commercially important fishes live for several years and present the problem of overlapping generations, as well as problems of observation and sampling in aquatic environments. Birds have been much observed because they are aesthetically attractive animals; some are of economic importance and have been studied almost as intensively as some insects but, again, there is often the problem of sorting out overlapping generations. Among mammals, small rodents such as voles and field-mice have been subjects of many studies, but present problems because of their retiring habits. There are some data for large animals such as African game animals and whales; the desire to conserve rare and interesting species has stimulated observations and experiments in 'management' of their habitats and populations. Monitoring the populations of large animals can reveal changes in ecosystems so that it is possible to take remedial action before there are catastrophic changes in the community structure.

What about plants? There have been studies similar to those on the population dynamics of animals but these are comparatively rare, except for some long-term studies of phytoplankton. The approach to crop plants has usually been in terms of biomass rather than in terms of numbers; there has been much more work on limiting factors, such as nutrients, than on collecting data for life tables.

The organization of Block B

1   There is a Pretest, based on S100*, Unit 20. The answers to this indicate the principles that are taken as prerequisites for this Block.

2   After reading the Introduction and doing the Pretest, you will need the prescribed book *Insect Population Ecology*, called *IPE* in the rest of this text. *IPE* and the parts of this text associated with it are equivalent to two Units (Units 6 and 7). This text includes separate study comments for each chapter of *IPE*, directing your attention to the important points made there. Since some students may feel unprepared for some of the mathematical treatment in *IPE*, there are Sections in Unit 7 on logarithms and on graphs; there are references to Sections of *Mathematics for the Foundation Course in Science*** (called

* The Open University (1971) S100 *Science: A Foundation Course*, The Open University Press.

** The Open University (1971) S100M *Mathematics for the Foundation Course in Science*, The Open University Press.

4

MAFS here) which also deal with these topics. You are advised to work through *IPE* steadily from beginning to end, spreading your study over the two weeks allocated for Units 6 and 7.

3   Having studied *IPE* and the parts of this text associated with it, you should go on to Units 8 and 9. These form a single cor plex, there is a single Study Guide (at the beginning of Unit 8) and the SAQs and references for both Units are given at the end of Unit 9. Units 8 and 9 are concerned with organisms other than insects, but the approach to population dynamics developed in *IPE* is applied to these other organisms.

4   The study of population dynamics raises a number of problems of experimental design, and we expect that you will need to refer to parts of *The Analysis of Biological Experiments**. Units 8 and 9 are both comparatively short in order to leave you time to work on *ABE*.

You will need log tables and/or a slide rule and graph paper to tackle exercises, ITQs and SAQs in this Block.

Recommended reading for Block B

den Boer, P. J. and Gradwell, G. R. (ed.) (1971) *Dynamics of Populations*, Centre for Agricultural Publishing and Documentation, Wageningen.

Lack, D. L. (1966) *Population Studies of Birds*, Clarendon Press.

Solomon, M. E. (1969) *Population Dynamics*, Arnold. (This is a prescribed text for S100, Unit 20.)

Varley, G. C., Gradwell, G. R., and Hassell, M. P. (1973) *Insect Population Ecology*, Blackwell Scientific Publications. (This is a prescribed text for this Block.)

Watson, A. (ed.) (1970) *Animal Populations in Relation to Their Food Resources*, Blackwell Scientific Publications.

Wynne-Edwards, V. C. (1962) *Animal Dispersion in Relation to Social Behaviour*, Oliver and Boyd.

Pretest questions

PTQ 1   Define (a) fecundity and birth rate; (b) mortality and death rate.

PTQ 2   What is the relationship between mortality and survival?

PTQ 3   In a stable population, what is the relationship between pre-reproductive mortality and fecundity?

PTQ 4   Give one example of each of the following:
(a)   animals with discrete generations;
(b)   animals with overlapping generations;
(c)   organisms with exponential rates of increase.

PTQ 5   Define (a) *k*-value; (b) key factor; (c) density dependent factor.

PTQ 6   Define territory, and give an example of an animal with territorial behaviour.

PTQ 7   What are specific parasites?

---

* The Open University (1974) S321/S323 ABE *The Analysis of Biological Experiments*. This forms part of the supplementary material for this Course. It is referred to as *ABE* in the text.

Pretest answers

PTQ 1 (a) Fecundity is defined in different ways by different authors. In S100, Unit 20, Section 20.3, it was defined as the average number of fertilized eggs produced in her lifetime by a female of a given species. Sometimes the term is applied to the number of eggs or young produced by the female in one breeding cycle. Birth rate is the term for fecundity applied to mammals, and usually means the number of young produced per female per unit time.

(b) Mortality and death rate are often taken as synonymous, and usually defined as in S100, Unit 20, Section 20.3, as the percentage of the population dying during a given time or over a given part of the life cycle.

PTQ 2 Percentage mortality subtracted from 100 gives percentage survival for that period. When the numbers of survivors at various times of an age group or family or cohort (or the percentage of the original numbers that survive) are plotted against time, the curve is called a survivorship curve (e.g. see S100, Unit 20, Fig. 6).

PTQ 3 For a species with equal numbers of males and females, the total pre-reproductive mortality in a stable population should equal the fecundity minus 2: this means that on average only two of the young survive, exactly replacing their parents (see S100, Unit 20, Section 20.3).

PTQ 4 (a) Winter moth and many other insects have discrete generations, that is, the parents die before any of their progeny become adults.

(b) Man has overlapping generations, so have cats and other mammals.

(c) Bacteria in cultures, phytoplankton organisms such as diatoms in lakes in spring, and yeast cells in culture show exponential increase for limited periods of time (see S100, Unit 20, Section 20.3.1). Usually, in cultures, the rate falls off as the numbers approach a maximum; this gives the logistic curve. The calculation of the instantaneous rate of increase of a population is based on similar reasoning to the calculation of instantaneous growth rates (S323, Unit 3, Appendix 4) and the same equation applies:

$N_T = N_t \cdot e^{u(T-t)}$, where N_T and N_t are the numbers in the population at times T and t (and T is later than t), e is the base of natural logarithms, and u is the instantaneous rate of increase (and $U = 100u =$ per cent rate of increase). Note that exactly the same expression can be used to obtain the instantaneous mortality rate Z for a cohort or age group or family:

$N_T = N_t \cdot e^{-z(T-t)}$ and $Z = 100z =$ instantaneous death rate as a percentage.

PTQ 5 (a) k-values are logarithmic measures of the killing power of mortality factors (see S100, Unit 20, Section 20.4.1).

(b) The key factor is the mortality factor which accounts for most of the change in total mortality from generation to generation.

(c) A density dependent factor is one whose effect on a population is related to the size of that population so that the effect is proportionally greater when the population is larger.

These three terms are defined and discussed further in *IPE*.

PTQ 6 Territorial behaviour occurs when an individual (or pair or group of individuals) display typical 'aggressive' behaviour patterns towards other individuals of that species which enter an area which is the territory; generally the intruding individuals retreat from the area rapidly (see S100, Unit 20, Section 20.4). Pairs of adult tawny owls occupy territories in woodlands and nest in, and obtain all their food from, the area that they occupy and defend from other tawny owls (see S100, Unit 20, Appendix 1). Many mammal, fish and bird species display territorial behaviour, but the territories may play different parts in the lives of different species.

PTQ 7 Specific parasites are parasites that attack one host species only (or a group of a few closely related host species). The term is applied by entomologists to include 'parasitoids', insects such as the wasps *Nemeritis* and *Pseudeucoila* whose larvae actually kill and consume the host insect (see S100, Unit 20, Section 20.4.2).

Units 6 and 7

Insect populations

Contents

Study guide to Units 6 and 7

These two Units rely on the prescribed book *Insect Population Ecology* (*IPE*) to present the main hypotheses and facts about insect populations. You are expected to read the whole of this book unless you are very short of time.

The Introduction to Unit 6 comments on the mathematics in *IPE*. Students who feel happy about their ability to use logarithms and interpret graphs should continue with Unit 6 and start to read *IPE*. Students who feel doubtful about their mathematical ability should turn to Sections 7.0 and 7.1; in these you are advised to read parts of *Mathematics for the Science Foundation Course* (*MAFS*) which give basic information about logarithms and graphs. There are some exercises associated with Sections 7.0 and 7.1 to give you confidence that you will be able to extract the essential features of the mathematical arguments in *IPE* by examining the graphs. The mathematics are a tool for helping to investigate the biological phenomena of population change and stability, and you should concentrate on the biological significance of the relationships revealed by mathematical analysis.

Sections 6.1 to 6.9 inclusive consist of study comments on the nine chapters of *IPE*. You are advised to read the appropriate comments before you read each chapter so that you recognize the important points as you come to them. *IPE* includes exercises based on each chapter. You are advised to omit some of these (they are listed under the chapter comments) because they are very time-consuming unless you have access to a computer; you should attempt the others treating them as a type of SAQ. Section 7.2 gives detailed answers and comments on the recommended *IPE* exercises. There are SAQs based on *IPE* and answers to them at the end of Unit 7. You could attempt the *IPE* exercises and the SAQs when you have finished each chapter, or you could leave them until you have finished the whole book.

IPE includes a large number of references; none is given in this text. *IPE* also includes a glossary of terms to which you should refer when necessary.

Objectives

After studying Units 6 and 7 and *Insect Population Ecology*, students should be able to:

1 Define correctly and distinguish between true and false statements concerning each of the terms and principles listed in Table A1.

(Tested in *IPE* and in SAQs 1, 2, 3, 4, 6, 7, 8 and 9.)

2 Given appropriate information:

(a) Draw up and analyse life tables and calculate or compare k-values and reproductive rates.

(Tested in *IPE* and in SAQ 2.)

(b) Identify key factors, regulating factors, density dependent factors, density independent factors and inverse density dependent factors.

(Tested in *IPE* and SAQs 3 and 7.)

(c) Calculate areas of discovery (for parasites).

(Tested in *IPE*.)

(d) Evaluate effects of weather or a named parasite on a population.

(Tested in *IPE* and SAQs 6, 7 and 8.)

(e) Predict which (if either) of a pair of competing species will survive and which will become extinct.

(Tested in *IPE* and SAQs 4 and 5.)

(f) State the consequences of given ways of altering the level of a population.

(Tested in *IPE* and SAQs 2, 4, 5, 7 and 8.)

3   Apply the principles developed in these Units and the prescribed text to make hypotheses or design experiments or draw conclusions from data relating to situations or to species not treated in the Units.

(Tested in *IPE* and SAQ 8.)

4   Assess published papers in the light of the principles developed in these Units distinguishing between valid and invalid methods of analysis and conclusions.

(Tested in *IPE*.)

Table A1

List of scientific terms, concepts and principles

Introduced in *IPE*	Page No.
	IPE
total, partial and cumulative population curves	3, 4*
generation histograms and curves	4*
k-values and key factor analysis	7*
competition	11
intrinsic rate of increase (natality and fecundity)	12
Verhulst-Pearl equation and logistic curves	13*
density dependence	18*
density independence	18*
inverse density dependence	19*
regulated population	19
logarithmic reproduction curves	22*
scramble and contest competition	25*
competitive exclusion	37*
Lotka-Volterra equations, interspecific competition	39
use of population models to predict population levels and changes	57
area of discovery of parasites	59*
delayed density dependence	65*
functional response	67
mutual interference constant of parasites	69*
Quest constant of parasites	69*
determination of population change	87
life tables	94*
integrated control of pests	131
biological control of pests	154

* The term appears in the Glossary of *IPE*.

Table A2

Units and Sections of Open University Courses taken as prerequisites

Course	Unit	Section	Objectives
S100	20	20.3	6
		20.4	7, 8, 9
		20.6	12
S100	21	21.5	2, 10
S22–	*IS***	Arthropoda A and B, D1	
	11	11.4	4

** The Open University (1972) S22– *Invertebrate Survey*. This forms part of the supplementary material for the S22– Course.

6.0 The mathematics of population dynamics

It will scarcely come as a surprise that the study of numbers of organisms can involve mathematical equations which look formidable to the non-mathematician. Some students may have studied calculus; they should have no difficulty in following the mathematics in *IPE*.

Students with little experience of mathematics should not worry because they are not able to understand and follow all the mathematical arguments in the text of *IPE*. The relationships between the variables studied are shown by graphs as well as in equations; all students should be able to interpret these graphs and thus see how change in one variable affects another. It is important to try to understand the graphs, but there is no need to despair if a few of them remain incomprehensible. Much of the interest of the topic lies in comparing predictions using 'models' based on different theoretical approaches with actual observations, and you should look very carefully at Figures in which such comparisons are presented. You can then judge for yourself how well the models conform to reality.

The authors of *IPE* are all biologists, and they use mathematics as a tool for exploring and clarifying the biology of insect populations. It is important that mathematical relationships should be assessed in terms of their biological implications, since it is possible to produce correlations and formulae which look splendid in print but which are nonsense biologically. You should, therefore, try to understand what Figures and formulae really mean in terms of observations that could be made in the field or in laboratories. You can then decide whether it is possible to test the relationships and judge the outcome of such tests.

If you feel unhappy about the use of logarithms or about the interpretation of graphs, turn now to Sections 7.0 and 7.1, which provide very simple guides to some of these problem areas and refer you to relevant Sections of *MAFS*.

6.1 IPE Chapter 1 Expressing population changes

Study comment The basic information for the study of population dynamics is a set of census figures—the numbers of individuals in the population at different times. Insects typically have life histories with the pattern: egg → larva → pupa → adult. In this chapter, the methods of displaying information about numbers is discussed; different types of graphical presentation can give very different impressions of how numbers change, so it is important to realize the differences between: total population curves; partial population curves; cumulative population curves; generation histograms; generation curves. Mortality and survival are two related aspects of studying how numbers change, illustrated in *IPE* Table I. The *k-value* or killing power of a mortality factor is defined in Section 1.6; it is very important that you should understand how this value is calculated and what it means because *k*-values are used frequently in the rest of the book (and also in Units 8 and 9).

graphical display of information

k-value

There are two exercises on Chapter 1—see Section 7.2, p. 14, for comments on these.

You could attempt SAQs 1 and 2 now. (see p. 25).

6.2 IPE Chapter 2 Density dependent processes affecting cultures of single species

Study comment The term *competition* is defined precisely in Section 2.2, and the rest of the chapter is a study of competition between individuals of the same species (intraspecific competition). The term 'intrinsic rate of natural increase' is introduced in Section 2.3, where the *logistic curve* (or Verhulst-Pearl equation) is explained as the result of actual rate of increase per individual falling as the population numbers rise. There is no need for you to follow the mathematical reasoning, but you should be able to explain in simple language the shape of curves such as those in Figure 2.2. In Section 2.4, possible relationships between percentage mortality and population density are

intraspecific competition

5

defined. This is a very important Section and you should work through it carefully, making sure that you understand Figure 2.6 and the terms: density dependent; density independent; inverse density dependent; regulated population.

relationships between density and mortality

Reproduction curves are used in Section 2.5 to explain how density dependent relationships may lead to different patterns of fluctuations in population numbers as a result of over, under and exact compensation. Section 2.6 defines two contrasted types of density dependent relationship: contest and scramble competition. Nicholson's well-known observations on populations of the sheep blow-fly *Lucilia* are reinterpreted by using *k*-values and reproduction curves; the changes in numbers are explained in terms of scramble competition.

reproduction curves

There are nine exercises on Chapter 2. Compare the first one with your Home Experiment Notes (*Drosophila* and *Daphnia*); see Section 7.2 for comments on the others. Note that in Exercise 2.6 you should start by plotting the logarithmic *survival* curve and then consider how to treat the reproduction curves.

You could attempt SAQ 3 now.

6.3 IPE Chapter 3 Competition between species for a limited resource

Study comment Most of this chapter deals with laboratory experiments, many on flour beetles which can be reared under strictly controlled environmental conditions. Section 3.3 provides background information. The basic theory, as propounded by Lotka and Volterra, is given in Section 3.4 both in mathematical terms and in the form of simple diagrams (Fig. 3.4). You should read the last two paragraphs of this Section, but could omit the rest if you find the mathematics difficult. You should read rapidly through the rest of the chapter, bearing in mind the biological implications of the Lotka-Volterra equations—that when two species with very similar biological properties and requirements come into the same environment, one will usually eliminate the other (this is competitive exclusion), but under some circumstances the two will coexist. The first paragraph of Section 3.9 sums up the conditions under which competitive exclusion or coexistence occur.

Lotka–Volterra equations

competitive exclusion

There are two exercises on Chapter 3; one is about a practical experiment and you should omit the second.

You could attempt SAQs 4 and 5 now.

6.4 IPE Chapter 4 Parasites and predators

Study comment In biological control it is desirable to be able to predict the outcome of adding to a large host population a specific predator or parasite (effectively the same as a predator) so the mathematical theory of population regulation owes much to applied entomologists. Thompson's model (Section 4.3) fits the early stages of increase in observed populations of introduced parasites, but predicts extinction of host and parasite (which does not usually occur). Nicholson's models (Section 4.4) assume that parasites have a constant 'area of discovery' and search at random for hosts; he believed in density dependent regulation of populations and was interested in equilibrium states. The mathematical expressions are used to calculate curves which can be compared with observed changes in numbers (e.g. Fig. 4.4). The term 'delayed density dependent factor' is defined in this Section, and 'functional response' is defined in Section 4.5, which shows how observations on parasites (or predators) in action leads to modification of Nicholson's models. The importance of studying behaviour is stressed again in Section 4.6 with the introduction of Quest Theory and the 'mutual interference constant'. Note in Figure 4.9 how changing the value of *m* alters the shape of population curves of both host and parasite, leading to extinction, or to stable oscillations, or to a steady state. Section 4.7 illustrates some problems in applying models derived from parasites to predators, and Section 4.8 gives evidence that search by parasites (or predators) is probably not random. They can aggregate or concentrate at local high population densities.

area of discovery

population regulation of hosts and parasites

Quest theory

There are four exercises on Chapter 4; see Section 7.2 for comments on 4.2 and 4.3. You should omit Exercise 4.4 unless you have access to a computer.

You could attempt SAQ 6 now.

6.5 IPE Chapter 5 Climate and weather

Study comment Section 5.4 discusses various ways in which weather and climate can affect insects; Section 5.5 gives examples of population change related to weather. The view of Andrewartha and Birch (1954)—that it is unnecessary to postulate density dependent factors to account for population 'control'—is criticized. You should pay special attention to the discussion related to Figure 5.8, which emphasizes the difference between factors that *determine population change* and factors which *regulate population level*. Weather can act as a density independent catastrophic factor determining population change, but regulating factors must have some density dependent effect. The loose use of the word 'control' to include both types of effect on population numbers has led to great confusion; this is why 'population control' is not used as a scientific term in *IPE* or in this Course.

determination of population change by weather

There is one exercise on Chapter 5; see Section 7.2 for comments.

You could attempt SAQ 7 now.

6.6 IPE Chapter 6 Life tables and their use in population studies

Study comment This chapter introduces 'population modelling', the basic technique for predicting the outcome of alterations in mortality factors. The first step—the construction of life tables for the insect and its parasites—is illustrated using data for the knapweed gall-fly, *Urophora jaceana*, and four parasites; you should work carefully through Sections 6.4 to 6.8, checking each calculation (these are simple arithmetic, using logarithms for some columns). Section 6.9 shows how the values in the life tables can be used to determine the relative effects on the population level of *Urophora* of a very large indiscriminate mortality (which also affects the parasites) and of a much lower mortality due to parasitism. The conclusion that abolition of the indiscriminate mortality leads to an increase in the number of parasites and consequent reduction in the number of adult hosts is of very great practical interest because many insecticides act as indiscriminate mortality factors. A basic concept in this chapter is the sequential action of mortality factors; this is the basis of 'key factor analysis', the use of k-values.

constructing knapweed gall-fly life tables

simple models using k-values

There are four exercises on Chapter 6; you will probably not be able to do the first. See Section 7.2 for comments on exercise 6.3; you should omit 6.4 unless you have plenty of time, because it will take at least half an hour.

6.7 Chapter 7 Interpretation of winter moth life tables

Study comment The winter moth, *Operophtera brumata*, has been studied for more than twenty years in Wytham Wood, near Oxford; some of the data collected are analysed in this chapter by construction of life tables and calculation of k-values. The general story should be familiar to you (S100, Unit 20, Sections 20.3 and 20.4), and you should be able to work fairly fast through Sections 7.3 to 7.8. Section 7.9 describes how a population model was constructed for winter moth and its parasite (do not worry about the details in Table 7.5); this model fitted reasonably well with actual numbers, as demonstrated in Figure 7.8. The model is used in Section 7.10 to illustrate the probable effects of altering mortality factors, e.g. by use of chemical sprays. When studying Figure 7.9, you should realize that *reduction* in a k-value means following the lines showing mean population density towards the *left*; the value of the slope of k_5 in the present model is 0.35 (the extreme right-hand value).

key factor analysis

construction of models with predictive value from winter moth data

There are two exercises on Chapter 7; exercise 7.1 puts you in the situation of starting with new data so it will take at least 45 minutes and you are advised to omit it and to tackle 7.2, where the data are already tabulated and you are required to construct and interpret graphs. See Section 7.2 for comments on exercise 7.2.

6.8 IPE Chapter 8 Population changes of some forest insects

Study comment If you are very short of time, you could omit this chapter. Sections 8.2 to 8.6 discuss examples of forest pests and the limited amount of information available on the factors which lead to outbreaks. Some of these species show cyclic changes in numbers which must be the result of delayed density dependent factors, but in no case is there absolutely convincing evidence of what the operative factor really is. In Section 8.7, the potential usefulness of k-values in constructing biologically realistic models is stressed; from such models could come accurate predictions both of population changes and of population levels under different conditions.

insect population cycles

There are three exercises on Chapter 8; these are all difficult and you are advised not to attempt them unless you are very interested (or have access to a computer) No comments are given here.

6.9 IPE Chapter 9 Biological control

See TV programme 'Biological control for examples.

Study comment This chapter shows how the theory of insect population dynamics relates to practical situations faced by applied entomologists. Notice the word 'control' is used here in the way it occurs in the literature of economic entomology. Sections 9.2 to 9.4 are concerned with examples illustrating various aspects of biological control. Section 9.5 discusses the use of mathematical models to select desirable properties of agents for biological control and to predict the outcome of introductions of agents. Section 9.6 considers some recent developments in biological control and problems presented by the use of insecticides.

models and desirable properties of agents for biological control

There is one exercise on Chapter 9; it is very time-consuming unless you have access to a computer. See Section 7.2 for comments; you are advised to read these even if you do not try the exercise.

You could attempt SAQs 8 and 9 now.

6.10 Summary

The first stage of studying population numbers is to collect data; then those data must be analysed and this raises the question of how to display them. Different types of graphical presentation may give very different impressions of what is happening in the population. Populations change as a result of natality (the production of offspring, related to fecundity), mortality (the deaths of individuals from any of a variety of causes), and the balance between immigration and emigration. In the study of insect populations, movements are often ignored; for mobile pests, such as locusts, the population is followed by the observers so the locality changes but not the individuals (except as a result of natality and mortality). Mortality is most conveniently expressed as k-value (killing power on a logarithmic scale); this is defined in *IPE*, Chapter 1.

natality and mortality

Competition between individuals occurs as a result of 'resources' being in short supply; the effects are density dependent, i.e. they become greater when the individuals are more crowded. The logistic curve (based on the Verhulst-Pearl equation) is an example of density dependence affecting recruitment. When an insect population with synchronized generations is affected by a density dependent mortality, the subsequent population changes depend on the reproductive rate (F) and on whether competition is weak, moderate or severe (this is measured as k-value); with moderate competition which under-compensates or slightly over-compensates for population change, the numbers reach a stable level, i.e. the population is regulated. The terms 'competition', 'density dependence', 'inverse density dependence', 'density independence' and 'regulated population' are all defined in *IPE*, Chapter 2.

population regulation

When two species with very smilar biological properties and requirements come into the same environment, one usually eliminates the other; this is called 'competitive exclusion' and is explained by the equations of Lotka and Volterra. When, for each of the two species, the effects of intraspecific competition are

competitive exclusion

more severe than those of interspecific competition, the two will coexist. Some observed examples of coexistence can be explained by assuming that there is a non-linear relationship between population density and reproductive potential; this may explain why coexistence of closely related species is more widespread than would be expected from theory.

coexistence

Three sets of hypotheses aimed at explaining interactions between insect hosts and their specific parasites (or parasitoids) are discussed in *IPE*, Chapter 4. A model which includes functions describing the egg production, the searching behaviour and the mutual interference between the adult parasites, as well as the egg production of the host, gives predictions of population changes of both hosts and parasites that are close to observed values. Chapter 4 illustrates the importance of observing the insects in action rather than approaching the topic purely from theory.

host–parasite interactions

Weather is defined as the changing hour-to-hour measurements of temperature, humidity, wind and rain, whereas climate represents the long-term averages of these measurements. Climate may restrict the range of insect species. Weather can act as a density independent or catastrophic factor, and determines population change in many insects. It is not a density dependent factor and cannot regulate population level. Confusion about 'control' of population numbers by weather is the result of different uses of the word 'control', so it is best avoided. The views of Andrewartha and Birch (1954)—that density dependent factors are often not involved in population regulation of insects—is criticized and refuted in *IPE*, Chapter 5.

effects of weather

When different mortality factors act in sequence on an insect population, the setting out of life tables and the calculation of k-values can lead to the development of models from which can be predicted the outcome of changes in these k-values. With a complete life table for one generation only, it is possible to deduce how the level of host and parasite populations will be altered by changes in the different mortality factors; with observations extended over many generations, the predictions become more precise because the parameters determining how the functions in the model should vary become more realistic. The use of a single life table to set up a model is described in *IPE*, Chapter 6, using data for the knapweed gallfly, *Urophora jaceana*. In *IPE*, Chapter 7, a long series of observations on winter moth, *Operophtera brumata*, are analysed to show how change in the level of the population at Wytham is determined by 'winter disappearance', a density independent factor which probably includes weather effects, whereas the population is regulated by density dependent 'pupal predation'. The winter moth population model has been used to explain the sequence of events in Canada, where this insect recently became an important pest and is now regulated at a low level as a result of the introduction of two European parasites.

life tables

key factor analysis

population models

Cyclical changes in populations of forest insects are discussed in *IPE*, Chapter 8; because necessary figures are not available, it is not possible to analyse any of these completely. The comparison shows clearly how the different approaches of investigators lead to the collection of different sorts of information and so limit the types of analyses that are possible.

The practical reason for studying and analysing the factors that affect insect populations is the hope of 'controlling' insect pests with great efficiency and at little cost. *IPE*, Chapter 9, discusses examples of 'biological control', some successful and others less so. The advantage of this system is that the only expense is the initial one of search for 'agents', their careful testing and their introduction; the insects then regulate the pest population indefinitely at no further cost. The ideas introduced earlier in *IPE* are applied in this final chapter to try to explain the success or otherwise of the examples given, and thus to set out the properties that a successful 'agent' should have and the investigations needed before a system of 'integrated control' (a combination of use of insecticides, parasites and predators) is applied. Application of insecticides, acting as an indiscriminate mortality factor on pest and predators, often results in raising the level about which the pest population fluctuates. The need to understand the biology of the pest and its potential controlling agents is clearly illustrated.

biological control

integrated control

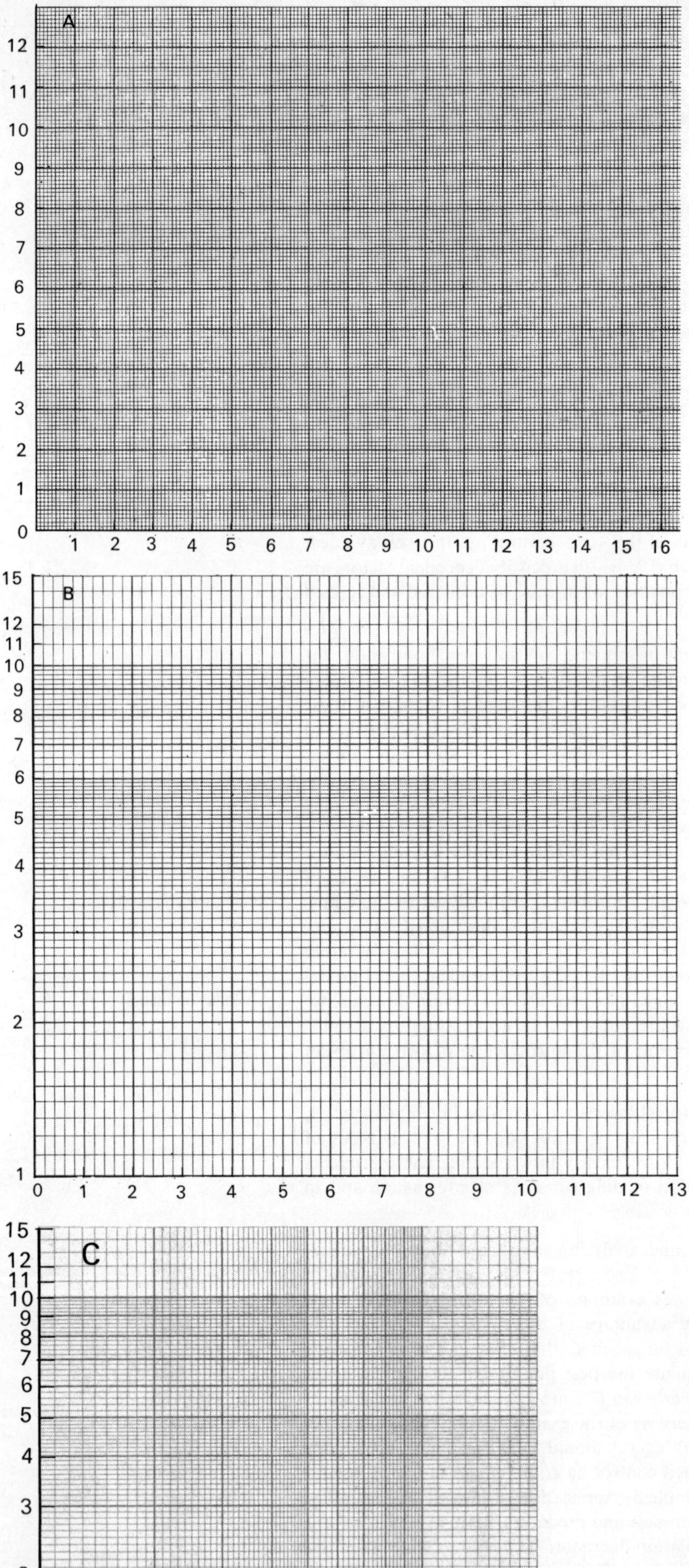

Figure 1 Types of graph paper: A, arithmetic paper; B, semi-log paper; C, log-log paper.

10

7.0 Logarithms

Read Section 1B of *MAFS* (probably you will find it helpful to read Section 1A first). Note that the description of logarithm tables based on Clark's *New Physical and Mathematical Tables* also applies to the *Science Data Book* (Tennent, 1971) which we have advised you to use.

Logarithms are a very useful tool for multiplying, dividing, obtaining square roots and so on (operations which you can also carry out on a slide-rule). They are extensively used in *IPE* in the form of k-values for mortality factors. The k-value is the difference between the logarithms of the population numbers before and after that mortality acts; you may wonder what sort of biological implication there is in this transformation of raw data.

TABLE 1

Population number		Logarithm of population number	k-value
stage 1	2000		
stage 2	1000		
stage 3	750		
stage 4	75		
stage 5	2		

Consider the example in Table 1 which shows how the numbers in a population varied in a series of stages of the life history of an imaginary organism. From these figures, try to answer the following questions:

ITQ 1 (a) Between what stages did the greatest number of individuals die?

(b) Between what stages did the greatest proportion of individuals (alive at the beginning of that stage) die?

(c) If you wished to increase the population alive at stage 5 and were allowed to halve the percentage mortality in just one of the intervals between stages, which interval would you choose in order to have the maximum number alive at stage 5?

Now, using the blank columns in Table 1, transform the population numbers into common logarithms (fill in the column in Table 1) and work out the k-values by subtracting each logarithm from the one above it. See the margin on p. 13 for logarithms and k-values (Table 2). Now answer question (c) again.

Read the answer to ITQ 1 on p. 26.

Look at the k-values: that for stages 4 to 5 (1.57) is the highest, followed by the k-values for stages 3 to 4 (1.00), stages 1 to 2 (0.3) and stages 2 to 3 (0.13). So this way of setting out population figures makes it easy to see the possible effects of altering mortality at different stages (times, ages) of the lives of the individuals.

Notice that halving the population gives a k-value of 0.3, and reducing the population to one-tenth gives a k-value of 1.00. This is true no matter what number of individuals you start with (confirm this by choosing numbers for yourself and calculating the k-values).

k-values and changes in population numbers

QUESTION What is the k-value associated with reducing the population to: (a) one-third; (b) one-quarter; (c) one-fifth; (d) one-thousandth; (e) one-hundredth.

ANSWER (a) 0.48; (b) 0.60; (c) 0.70; (d) 3.0; (3) 2.0. You may have noticed that these figures are each the logarithm of the denominator of the fraction (i.e. to divide by 3, subtract log 3 from the log number of individuals).

The use of logarithms makes it easy to compare sets of figures to see how mortality factors vary from year to year or from place to place.

7.1 The interpretation of graphs

Read Sections 3A, 3B and 3C of *MAFS*; these explain how the coordinates of a point can be described, how to calculate from the 'slope' of a graph the equation which expresses the relationship between the points on it, and how to construct a graph from given data. You will probably find Sections 3A and 3C very simple, but it is important that you should be able to carry out the sort of calculation illustrated in Section 3B.

The graphs discussed in *MAFS* are on arithmetic coordinates; in Block B you will meet many graphs drawn on coordinates which are logarithmic either in one direction (semi-log plots) or in both directions (log-log plots). Why are these plots used and what to they show?

arithmetic and logarithmic plots

Figure 1 consists of three blank sheets of graph paper; A is arithmetic graph paper, B is semi-log and C is log-log with just over one 'cycle' on each axis. Compare these carefully; note that the log scales start at 1 whereas the arithmetic scale starts at 0.

Figure 1 is on p. 10

(a)   Now draw a straight oblique line on the arithmetic paper, read off the coordinates at about six convenient well-spaced points. Plot these points on the semi-log paper and then on the log-log paper. What happens to your straight line on the other two plots?

(b)   Draw a straight line on the semi-log paper, read off the coordinates at about six convenient places. Plot these points on the arithmetic paper and on the log-log paper. What happens to your straight line on the other two plots?

(c)   Draw a straight line on the log-log paper, read off the coordinates at about six convenient places. Plot these points on the arithmetic paper and on the semi-log paper. What happens to your straight line on the other two plots?

Compare your graphs with those in Figure 2A, B and C. In the case shown:

(a)   The straight line on arithmetic paper becomes a convex curve on semi-log paper and a slightly concave curve on log-log paper.

(b)   The straight line on semi-log paper becomes a concave curve on arithmetic paper and a concave curve on log-log paper.

(c)   The straight line on log-log paper becomes a convex curve on semi-log paper and a slightly convex curve on arithmetic paper.

Note that there is more contrast between the shapes on semi-log paper and the other two than between these other two (arithmetic and log-log paper).

There are two very useful properties of curves plotted on log scales:

advantages of log plots

1   The logarithmic scale means that changes at the lower end (in the digits or tens) are made just as obvious as changes in the hundreds or thousands; an equal change in the scale means an equal change in proportion, i.e. one-third, one-half or whatever it may be; check this on your graph paper.

2   If the abscissa represents time on an arithmetic scale, then the slope of the curve represents the rate at which the variable plotted on the ordinate (log scale) changes.

There is no need to acquire semi-log or log-log graph paper (which are expensive); the same result is achieved by plotting the logarithms of the numbers on arithmetic scales. If you need to convince yourself of this, try it with the data you collected when plotting your lines.

Now read quickly through Section 5C of *MAFS*, 'Exponential growth and decay'. Note that if these curves were plotted on semi-log paper, they would become straight lines. In this Section of *MAFS*, the calculus notation is used, e.g.

calculus notation

$\frac{dy}{dt}$. Note that the d in this simply denotes a very small value of y or t. In population studies, the expression $\frac{dN}{dt}$ is frequently used with N = population numbers and t = time. The expression means the rate of change in numbers with time.

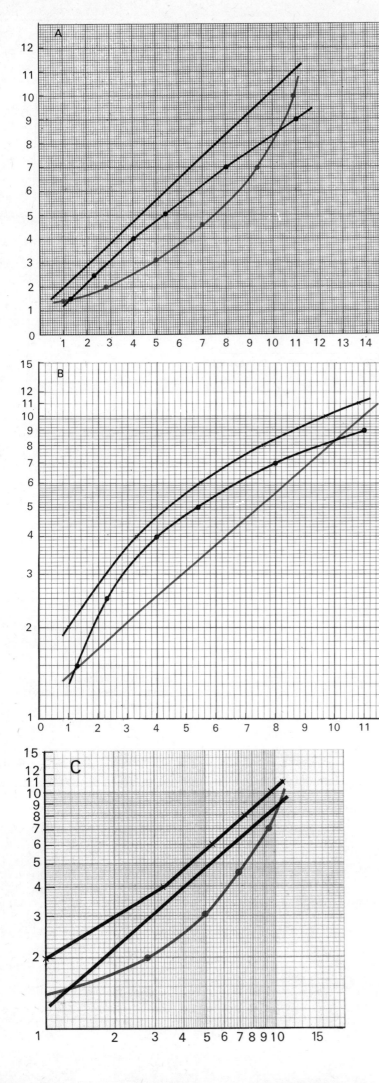

TABLE 2

Logarithm of population number	k-value
3.3010	
	0.3010
3.0000	
	0.1249
2.8751	
	1.0000
1.8751	
	1.5741
0.3010	

13

QUESTION   What is the change in numbers with time when $\dfrac{dN}{dt} = 0$?

ANSWER   There is no change, i.e. the population numbers remain constant (often called stable).

QUESTION   What would you expect to be the meaning of $\dfrac{dN_1}{dt} \times \dfrac{1}{N_1}$?

ANSWER   This means the rate of population change divided by the numbers in the population. The expression N_1 means the numbers in a particular population or the population of a particular species as compared with another population which would be N_2.

7.2   Comments on Exercises in IPE

Exercise 1.1

Figures extracted from Table 1.1

Successive percentage mortality	0	50	50	60	90	99 (overall)
k-value for the same interval	0	0.3	0.3	0.4	1.0	2.0 (sum)

The graph is shown as Figure 3.

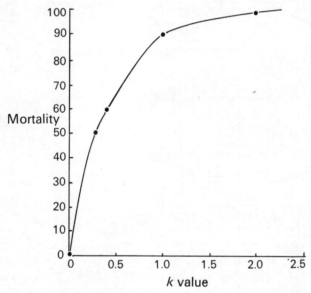

Figure 3 Successive percentage mortality (ordinate scale) plotted against k-value (abscissa scale). (Exercise 1.1)

Exercise 1.2

	Eggs	Instar 1	Instar 2	Instar 3	Instar 4	Prepupa	Pupa	Adult
Population	1000	576	343	172	93	51	26	11
Number dying		424	233	171	79	42	25	15
Percentage mortality		42.4	23.3	17.1	7.9	4.2	2.5	1.5
Successive percentage mortality		42.4	40.5	50	46	45	49	58
Successive percentage survival		57.6	59.5	50	54	55	51	42
Fraction surviving		0.58	0.6	0.5	0.54	0.55	0.51	0.42
Log population	3.0	2.76	2.54	2.24	1.97	1.71	1.42	1.04
k-value		0.24	0.12	0.30	0.27	0.26	0.29	0.38

(a)   The largest number died between egg and first instar;

(b)   The greatest fraction survived between first and second instars;

(c)   The greatest fraction died between pupa and adult;

(d)   0.38 (or 0.37 if you use logs to four figures).

14

Exercise 2.2

Tabulation of data from Figure 2.6A (plotted in Fig. 4)

Population	1	10	20	30	40	50	60	70	80	90	100
Percentage mortality	1	10	20	30	40	50	60	70	80	90	100
Log N before	0	1	1.3	1.48	1.6	1.7	1.78	1.85	1.90	1.95	2
Log N after	$\bar{1}$.99	0.95	1.2	1.32	1.38	1.4	1.38	1.32	1.20	0.95	0
k-value	0.01	0.05	0.1	0.16	0.22	0.3	0.4	0.55	0.7	1.0	2

Notice the very steep rise in mortality with increase of density; this is especially marked when the k-value is plotted against the logarithm of the population.

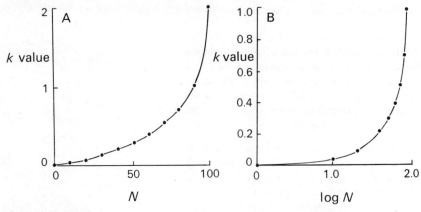

Figure 4 k-value plotted against population density N on arithmetic (A) and logarithmic (B) scales. (Exercise 2.2)

Exercise 2.3, 2.4, 2.5

Tabulate the data:

Number n	Percentage mortality	Number of survivors	Offspring produced (= number $n+1$) if		
			$F = 2$	$F = 3$	$F = 5$
1	1	0.99	2	6	10
10	10	9	18	27	45
20	20	16	32	48	80
30	30	21	42	63	105
40	40	24	48	72	120
50	50	25	50	75	125
60	60	24	48	72	120
70	70	21	42	63	105
80	80	16	32	48	80
90	90	9	18	27	45

The curves are plotted in Figure 5. Note that when the number rises above 100, all must become extinct. To answer the rest of the questions, start anywhere on

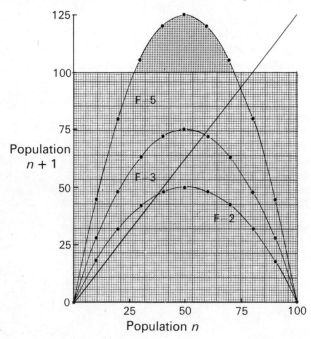

Figure 5 Reproduction curves plotted on arithmetic scales. The shaded area is equivalent to extinction (100 per cent mortality). (Exercises 2.3, 2.4, 2.5)

15

the abscissa and then proceed as in Figure 2.7C. With the curve for $F = 2$, the population is stable; with $F = 3$, the population will fluctuate about a stable level; but with $F = 5$, the population soon reaches a situation in which it becomes extinct. The key to recognizing whether or not a population is stable is to note the angle at which the reproduction curve cuts the diagonal; if this angle is acute, then the population is stable, but if it is obtuse, the population will be very unstable and may become extinct. If the reproduction curve is horizontal (or rising) where it intersects the diagonal, the population will rise to a stable level; otherwise there will be fluctuations because of over-compensation.

Exercise 2.6

Draw the logarithmic survival curve by plotting the log of the number of survivors (from data for Exercises 2.3 to 2.5) against the log of the population number n. Now plot, with the same scale on the ordinate, the log of the $n + 1$ population (when $F = 2$ from the same data). Compare the shape of this curve with the survival curve—they are identical, but the curve for population when $F = 2$ lies always 0.3 above the curve for population when $F = 1$ (i.e. the survival curve). Check that 0.3 is the log of 2. (See Fig. 6).

In fact, instead of drawing different reproduction curves for each value of F, it is possible to draw the survival curve and to transform this into any reproduction curve by *altering the values of the ordinate scale*. To transform the survival curve ($F = 1$ so that $\log F = 0$) into the reproduction curve $F = 2$, simply change the ordinate scale by subtracting 0.3 from each value. What value would you subtract to turn the $F = 1$ curve into the reproduction curve for $F = 5$? Answer given below.*

To discover whether the population would become stable at any particular rate of reproduction, it is necessary to draw the diagonal line which connects equal values of n and $n + 1$ (i.e. starts from 0, 0 and goes to 2.0, 2.0). For the survival curve, this line is always above the curve; this implies that, if the reproductive rate was one offspring for each individual, the population would gradually decline and become very small and eventually extinct. The diagonal using the ordinate values appropriate for when $F = 2$ crosses the curve at an acute angle and where it is horizontal; this rate of reproduction would give a stable population. The diagonal using the ordinate values for $F = 5$ crosses the curve at an obtuse angle and this population would soon become extinct. These three diagonals are, of course, parallel to each other.

To use the curve to answer questions (a) and (b) means trying to construct diagonals that will cut the survival curve at appropriate places; then the value at which this diagonal cuts the ordinate scale is the logarithm of the appropriate reproductive rate.

(a) For this, the diagonal must cut the survival curve at right angles; the way to find this point is to draw the tangent to the curve which would be at right angles to the diagonal. On Figure 6, this diagonal cuts the ordinate scale at -0.47 which is the logarithm of 3. So $F = 3$ is the highest value at which the population is likely to stabilize, but with marked fluctuations because of over-compensation.

(b) For this value, the diagonal must pass through the point where the horizontal line which is a tangent to the survival curve intersects with the vertical line through $\log n = 2$ (because mortality is 100 per cent for this value). The diagonal is drawn parallel to the other diagonals; in this case, it cuts the ordinate scale at -0.6, which is the logarithm of 4. So, for values of F above 4, the population will become extinct. Remember also that for $F = 1$ the population slowly declines and reproductive rates below 1 will lead to extinction.

You should note that the situation being studied is highly artificial, but resembles that of some insect populations where there are annual generations and the adults breed only once and then die. If adults breed more than once and generations overlap, the situation is much more complex.

---

* 0.7 (which is log 5).

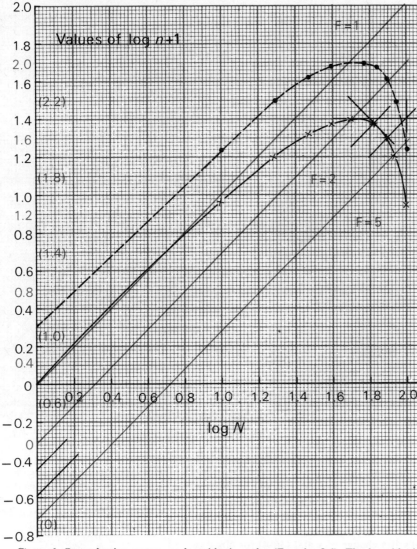

Figure 6 Reproduction curves on logarithmic scales (Exercise 2.6). The logarithmic survival curve and the scales for it are shown in black. The reproduction curve for $F = 2$, plotted on the same scales as the survival curve, is drawn in broken lines. The diagonals for $F = 1$, $F = 2$ and $F = 5$ are drawn in red and the appropriate ordinate scales for $F = 2$ and $F = 5$ are shown in red to the left and right respectively. The construction lines for answers (a) and (b) are shown in black and the diagonals for these are drawn only where they intersect the curve and where they intersect the ordinate scale.

Exercise 2.7

In this you repeat the operations given in Exercise 2.6, but starting with a different shape of density dependent mortality curve. The rules given for Exercise 2.6 should allow you to recognize the reproductive rates at which the populations will be stable, will fluctuate considerably and will become extinct.

Exercise 2.8

The values (plotted in Fig. 7A and B on p. 18) are:

| | *Cryptolestes* | | | *Cathartus* | |
Number of adults	Number of progeny	Progeny per adult	Number of adults	Number of progeny	Progeny per adult
4	101	25	4	96	24
8	180	22.5	8	287	36
16	276	17	16	331	21
32	427	13.4	32	333	10.5
64	411	6.4	64	434	6.7
128	473	3.7	128	392	3

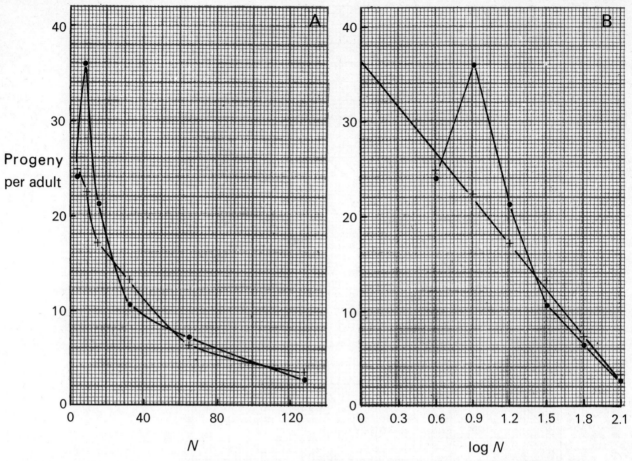

Note from Figures 7A and B that the two beetles give similar curves, but *Cathartus* has two points that diverge from the straight line of the plot of progeny per adult against log number of adults. To deduce the expression, note that the line cuts the ordinate at 36. The value for progeny per adult is 20 when the log number of adults is 1, and 4 when the log number of adults is 2; hence the slope of the line is (36–20) for abscissa change of 1 and (36–4) for abscissa change of 2 which gives 16 per change of 1. So the expression that fits the curve is: $F = 36 - 16 \log N$.

Figure 7 Production of offspring by beetles at different population densities (N) expressed arithmetically (A) and logarithmically (B). + —*Cryptolestes* ● —*Cathartus*. (Exercise 2.8)

Exercise 2.9

The tabulated data are:

Initial number of eggs N	$\log N$	log number of pupae ($\log P$)	k-value ($\log N - \log P$)
10	1	0.89	0.11
20	1.3	1.18	0.12
50	1.7	1.57	0.13
100	2	1.87	0.13
200	2.3	2.14	0.16
400	2.6	2.45	0.15
800	2.9	2.68	0.22
1600	3.2	2.59	0.61
3200	3.5	2.58	0.92
5000	3.7	2.51	1.19

(a)   From Figure 8, it is clear that this is contest competition (compare *IPE* Fig. 2.11).

(b)   The slope of the line is very gradual, so $b = 0.05$.

(c)   This slope is steep (0.6 on the ordinate for 0.5 on the abscissa) so 1.2 is the correct answer.

(d)   The maximum number of pupae produced was 477 from 25 g of food giving about 0.05 g per pupa.

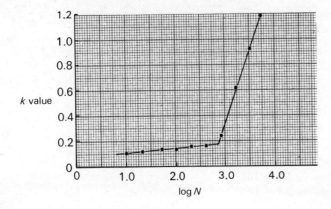

Figure 8 *k*-value plotted against log *N*.
(Exercise 2.9)

Exercise 4.2

Assume that the difference between *N* and *S* is the number of hosts found by parasites.

	Parasite density P	Host density N	Hosts found $N - S$	Hosts found per parasite $\dfrac{N-S}{P}$	Area of discovery (from formula given) a
	18	36	20.6	1.15	4.7×10^{-2}
	21	31	18.1	0.86	4.2×10^{-2}
	18	26	15	0.83	4.7×10^{-2}
A	15	22	10.7	0.71	4.4×10^{-2}
	11	23	8.7	0.79	4.3×10^{-2}
	9	29	10.7	1.2	5.1×10^{-2}
	11	37	13.5	1.22	4.2×10^{-2}
	40	1	0.99	0.025	5.4×10^{-2}
	40	5	4	0.1	3.5×10^{-2}
	40	25	17.57	0.44	3.0×10^{-2}
B	40	50	31.5	0.8	2.5×10^{-2}
	40	100	48.3	1.2	1.7×10^{-2}
	40	200	54	1.35	0.8×10^{-2}
	40	300	62	1.55	0.6×10^{-2}

To find the nature of the parasites' functional response, plot hosts found per parasite against host density (see Fig. 9).

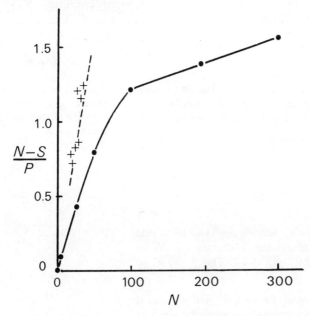

Figure 9 Numbers of hosts found per parasite against host density.
+—Experiment A ● Experiment B
(Exercise 4.2)

Figure 9 shows that as host density rises, the proportion of hosts parasitized rises, but above a host density of about 100 there is little improvement. The calculated area of discovery decreases as the number of hosts is increased. This suggests that some aspect of the parasite's behaviour means that it cannot deal with more than a small number of parasites each day, i.e. that there is a long 'handling time' for each host attacked. Compare Figure 9 with *IPE*, Figure 4.7A.

(b) The parasites in Experiment A seem to have been more effective than parasites operating over a similar range of host density in Experiment B. The main difference is that the parasite density in Experiment A was about one-quarter to one-half that in Experiment B, and this means that mutual interference should have been less in Experiment A and the population of hosts attacked by parasites was therefore higher. Compare the area of discovery of:

9 parasites provided with 29 hosts	5.1×10^{-2}
21 parasites	31 hosts	4.2×10^{-2}
40 parasites	25 hosts	3.0×10^{-2}
40 parasites	50 hosts	2.5×10^{-2}

Exercise 4.3

The areas of discovery were:

Parasite density (P)	1	2	4	8	16	32
Area of discovery (a)	0.17	0.096	0.064	0.049	0.025	0.017

The graphs are plotted in Figures 10A and B.

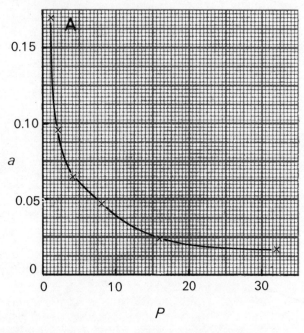

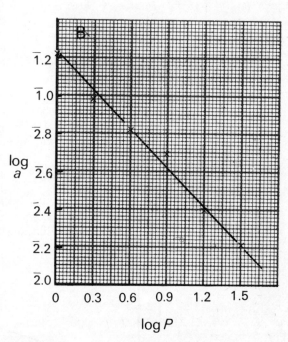

There is a straight-line relationship between log a and log P. The line crosses the ordinate at $\bar{1}.23$ (which is -0.77); the ordinate change corresponding to 1 on the abscissa is from $\bar{1}.23$ to $\bar{2}.56$ (which is 0.67 divisions). Thus the equation is:

$$\log a = -0.77 - 0.67 \log P$$

Figure 10 (A) Area of discovery a plotted against parasite density P and (B) log a plotted against log P. (Exercise 4.3)

Exercise 5.1

The k_1 and k_2 values are found by subtraction. See Figure 11A. Note that $F = 50$ as a constant from generation to generation. (Subtract log N_A of generation 1 from log N_E of generation 2 : 2.0 $-0.3 = 1.7$. This is log F, so F is 50.)

(a) Figure 11B shows a straight line relationship between k_2 and log N_L; it passes through the origin and the ordinate has a value of 1.6 when the abscissa is 2; the formula therefore is: $k_2 = 0.8 \log N_L$.

(b) The relationship described in the question can be seen in Figure 11A. Note that log N_E varies very little from year to year. This means that k_2 acting on N_L

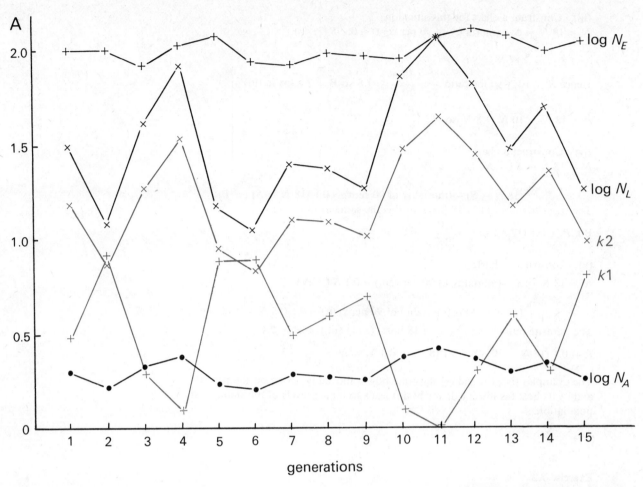

generations

leads to a great stability in adult numbers. Thus the random mortality k_1 determines the larval density N_L which determines k_2 through the strongly density dependent relationship shown in Figure 11B.

Exercise 6.2

The answer is the rest of the figures in Table 6.2 or 6.3. The calculations are explained in the text of *IPE*.

Exercise 6.3

(i) In the steady state, with $F = 18$, the ratio $\dfrac{N}{S} = 18$. Log 18 is 1.26.

Therefore $P = \dfrac{2.3}{0.25} \times 1.26 = 11.6$

$N = \dfrac{P}{17} = 0.68$ (because the parasite takes 17 of 18 N offspring, leaving one to survive to the next generation).

(ii) In a steady state, if $F = 18$ and is followed by 90 per cent mortality, then *Eurytoma* must prevent an increase of 1.8 (10 per cent of F).

$\dfrac{N}{S} = 1.8$ and the log of 1.8 is 0.26. Therefore $P = \dfrac{2.3}{0.25} \times 0.26 = 2.4$

To calculate N from P involves constructing a chart:

Parents   Larvae   Survivors   Next generation

N—($F = 18$)→ $18\,N$—(mortality of 90 per cent)→ $1.8\,N \rightarrow S\,(= N)$

$P\,(= 0.8\,N)$

so $N = \dfrac{P}{0.8} = 3$

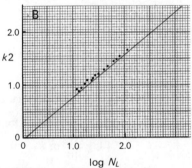

$k2$

$\log N_L$

Figure 11 (A) Generation curves (black) and k-values (red). The values on the ordinate scale are for log N (population numbers) and k-values. (B) Test of k_2 for density dependence. (Exercise 5.1)

(iii) Construct a chart for this situation:

$N \to 18\ N \to S$ —(mortality of 90 per cent)$\to 0.1\ S\ (= N)$

$18\ N - S\ (= P)$

Hence $S = 10\ N$ so the ratio $\dfrac{N}{S} = \dfrac{18\ N}{10\ N} = 1.8$ so $P = 2.4$ (as in (ii))

$P = 18\ N - 10\ N = 8\ N$, so $N = \dfrac{P}{8} = 0.3$

(iv) Construct a chart:

$N \to 18\ N \to S\ (= N)$

$(18\ N - S)$ —(mortality of 90 cent)$\to 0.1\ (18\ N - S)\ (= P)$

The parasite must kill 17/18 hosts so P is the same as in (i): 11.6

But $P = 0.1\ (17\ N)$ so $N = \dfrac{P}{1.7} = 6.8$

(v) Construct a chart:

$N \to 18\ N \to S$ —(mortality of 90 per cent)$\to 0.1\ S\ (= N)$

$(18\ N - S)$ —(mortality of 90 per cent)$\to 0.1\ (18\ N - S)\ (= P)$

The parasite must leave 10 out of 18 hosts (as in (iii)) so $P = 2.4$

$P = 0.1\ (18\ N - 10\ N) = 0.1\ (8\ N) = 0.8\ N$, so $N = \dfrac{P}{0.8} = 3$

The examples (iv), (ii) and (v) illustrate how a mortality acting on the parasite and/or its host (as insecticide might do) leads to higher levels of the stable host population.

Exercise 7.2

(a) In Figure 12A, clearly only k_1 varies in the same way as K.

(b) Looking at Figure 12B, C and D, only k_3 seems to vary in a density dependent way.

(c) The line for k_3 in Figure B starts at log $N_{L2} = 1$, which means $\dfrac{N_{L2}}{10}$. The rise is 0.8 on the ordinate for a change of 2 on the abscissa. So the formula is: $k_3 = 0.4 \log \dfrac{N_{L2}}{10}$ or $k_3 = 0.4\ (\log N_{L2} - 1)$. Recall that dividing a number by 10 is equivalent to subtracting 1 from the log of the number (see Section 7.1 and *MAFS*).

(d) In Figure 13A, the two variables are plotted against each other to give a straight line which has a slope of about 0.65 (this should be 0.6 from the relationship deduced under (c)).

(e) No. See the answer in *IPE*.

(f) The values calculated are:

Generation	1	2	3	4	5	6	7	8
$P\ (= P_A)$	250	262	473	115	173	168	279	70
log P	2.4	2.47	2.68	2.06	2.24	2.23	2.45	1.85
k_2	0.47	0.52	0.65	0.32	0.39	0.39	0.5	0.25
$a \times 10^{-3}$	4.3	4.0	3.2	6.4	5.2	5.4	4.1	8.2
log a	$\bar{3}.63$	$\bar{3}.6$	$\bar{3}.51$	$\bar{3}.81$	$\bar{3}.72$	$\bar{3}.73$	$\bar{3}.61$	$\bar{3}.91$

These figures are plotted in Figure 13 and give a straight line with a slope of 0.5 decrease on the ordinate for 1.0 increase on the abscissa. By extrapolation, the line would intersect with log $P = 0$ when log $a = \bar{2}.83$ (which is -1.17). So the formula is: log $a = -1.17 - 0.5$ log P.

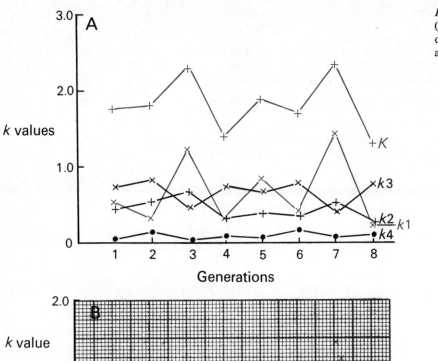

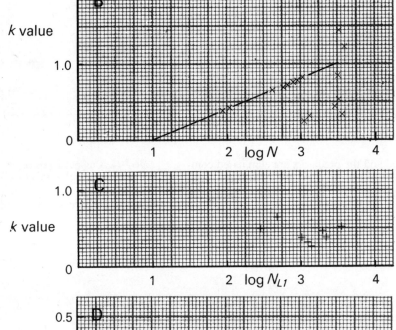

Figure 13 (below) (A) Log N_P against log N_{L2}. (Exercise 7.2). (B) Log area of discovery of parasite (*a*) against log parasite density (*P*). (Exercise 7.2)

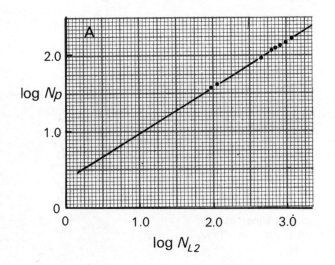

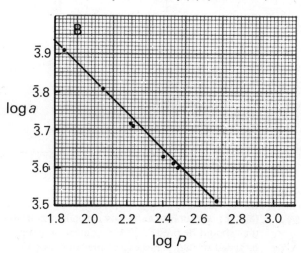

Exercise on Chapter 9

The values derived from a computer run starting with a pest density of 1000 and a parasite density of 30 and applying the formulae given in *IPE* with $F = 75$, $k_1 = 0.7$, $m = 0.52$ and $Q = 0.056$ gave a steady parasite density of 30.

(a)   The pest steady density is the antilog of 3.13 which is about 1350 pests.

(b)   Applying to this a 70 per cent kill of the pest just before parasitism in each generation gives the curve shown in Figure 14A. The pest steady density becomes the antilog of 2.34, which is about 220 pests (about 16 per cent of the previous value); the parasites stabilized at about 54.

(c)   If this kill also killed 70 per cent of the searching parasites in each generation, the curve would be as in Figure 14B. The pest steady density is antilog 2.37, which is about 235 pests.

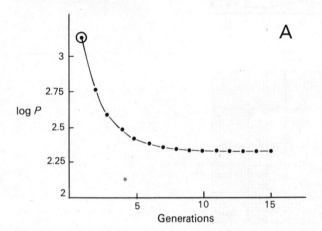

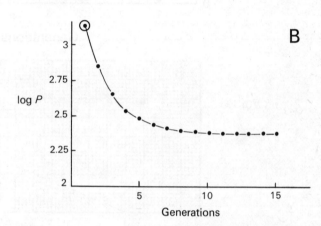

(d)   If 70 per cent of the pests are killed and the predators are eliminated completely, the pest population starts to undergo violent decreasing oscillations as illustrated in Figure 14C. Numbers at peak levels increase tenfold above the previous steady density.

Figure 14 Computer-generated curves for pest numbers (log P) following the manipulations of the population model described in Exercise 9. See text for explanation.

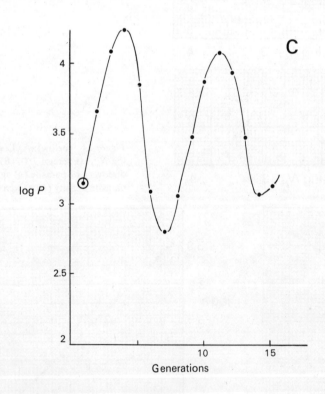

(e)   If predators act as density dependent mortality factors, it is important that they should be preserved. Any control programme directed at the pests should be timed so that parasites and predators are not also affected.

Self-assessment questions

SAQ 1 (*Objective 1*) Identify which of the terms (i) to (v) apply to the definitions A to E:

A The data for eggs, larvae, pupae or adults are shown separately as blocks of the appropriate height and arranged in order of generations.

B The total numbers of individuals of the species are plotted against time and the values are connected by a continuous line.

C The number of larvae, pupae or adults present are plotted against time and the values for each are connected by continuous lines.

D The total number of adults which have emerged is plotted against time and the values are connected by a continuous line.

E The total number of larvae in each generation is plotted against the order of generations and the points are connected by a continuous line.

Terms (i) total population curve; (ii) partial population curve; (iii) cumulative population curve; (iv) generation histogram; (v) generation curve.

SAQ 2 (*Objectives 1, 2 (a), 2 (f)*) Mark the statements (a)–(e) as either TRUE or FALSE.

(a) The sum of the k-values for a generation is equal to the generation mortality K.

(d) The sum of the successive percentage mortalities is equal to 100 per cent.

(c) When a population is in balance, the logarithmic increase log F must be greater than the generation mortality K.

(d) When population numbers are plotted on a logarithmic scale, decreases of 40 per cent in numbers will always appear as equal decreases measured vertically (on the ordinate scale).

(e) When population numbers vary by a hundredfold or more, they should be plotted on an arithmetic ordinate scale so that changes at the lower end of the scale are made visible.

SAQ 3 (*Objectives 1, 2 (b)*) Identify which of the terms (i)–(vii) apply to the definitions A–G.

A The birth rate minus the death rate under optimal conditions of those physicochemical factors which may affect the insects.

B The type of mortality where the proportion killed decreases as the population increases.

C The type of mortality where the proportion killed increases as the population increases.

D The type of mortality where the proportion killed bears no consistent relationship to the numbers in the population,

E The type of competition when the resource is shared equally between all the competing animals no matter how many or how few.

F The type of competition when the resource is shared in such a way that successful animals get all that they require no matter whether the population is large or small.

G A population whose density tends to return to an equilibrium value if it is displaced from this.

Terms (i) density dependent mortality; (ii) density independent mortality; (iii) intrinsic rate of natural increase; (iv) regulated population; (v) scramble competition; (vi) contest competition; (vii) inverse density dependent mortality.

SAQ 4 (*Objectives 1, 2 (e), 2 (f)*) Mark the statements (a)–(e) as either TRUE or FALSE.

(a) Interspecific competition occurs when organisms of one species exert a disadvantageous influence on organisms of another species because their more or less active demands exceed the immediate supply of their common resources.

(b) When two species compete, one will always eliminate the other; this is called competitive exclusion.

(c) When two species compete, the outcome depends on the balance between intraspecific and interspecific competition for each of the species.

(d) If one species has a more adverse effect on the other species than on itself, then competitive exclusion occurs.

(e) If both species have more adverse effects on members of their own species than on members of the other species, then the two species can coexist.

SAQ 5 (*Objective 2 (e), 2 (f)*) P and Q are two very similar species which compete for resources. What will be the outcome of competition between them under each of the conditions (a)–(c)?

(a) The increase in numbers of species P is inhibited by population densities (of both P and Q) of lower value than those which inhibit increase of species Q.

(b) The increase in numbers of species P is inhibited more severely by high densities of its own species than by high densities of the other species; the same statement applies to species Q.

(c) The population density at which species P inhibits increase of species Q is less than the population density at which species P inhibits its own increase; the same statement (with letters interchanged) applies to species Q.

The four possible outcomes are: (i) species P will eliminate species Q; (ii) species Q will eliminate species P; (iii) the two species will coexist in a stable way; (iv) the species may coexist for a short time but one will eventually eliminate the other. You should find it helpful to study *IPE*, Figure 3.4.

SAQ 6 (*Objectives 1, 2 (d)*) Consider the three models of host–parasite relationships discussed in *IPE*, Chapter 4: Thompson's (Section 4.3), Nicholson's (Section 4.4) and Hassell and Varley's (Section 4.6). Identify which of the statements (a) to (i) are essential parts of each of the three types of model.

(a) Encounters between parasites and hosts are random.

(b) Parasites act as delayed density dependent factors.

(c) An individual parasite lays one egg only in each host.

(d) A parasite searches for hosts over an area of discovery which is constant and characteristic of that species.

(e) A parasite population's rate of increase is limited only by the females' egg supplies.

(f) Parasites interact with each other in such a way that searching efficiency falls off with increase of parasite density.

(g) The rate at which parasites find hosts is limited by host density.

(h) A parasite searches for hosts over an area of discovery which is smaller when parasite numbers are higher.

(i) Parasites show a density dependent response to host distribution; their search is therefore not random.

SAQ 7 (*Objectives 1, 2 (b), 2 (d), 2 (f)*) Identify each of the statements (a)–(c) as either TRUE or FALSE.

(a) Climate is the seasonal means of conditions such as temperature, rainfall and light recorded over many years; it is greatly influenced by latitude and by altitude.

(b) The weather, as distinct from climate, affects survival of individual insects and may therefore have a great effect on population changes.

(c) In the absence of density dependent mortality, weather factors can regulate insect populations.

(d) In the presence of strong density dependent mortality, weather factors can act as key factors in determining population change.

(e) Weather factors are density independent and may be catastrophic factors.

SAQ 8 (*Objectives 1, 2 (d), 2 (f), 3*) List at least five attributes of insects which would make them suitable for use as agents of biological control. Give reasons for your choice.

SAQ 9 (*Objectives 1, 3*) The study of insect population dynamics has been carried out by workers in the field and in the laboratory. Make two lists: a list of those aspects of models which have been derived mostly from laboratory experiments; a list of those aspects of models which have been derived mainly from field studies.

ITQ answer

ITQ 1 Data from Table 1 and further calculations:

Stage	1	2	3	4	5
Population	2000	1000	750	75	2
Number of individuals dying before next stage	1000	250	675	73	
Percentage of individuals dying before next stage	50	25	90	97	
Number in population after altering percentages of individuals dying					
at stage 1 to 2	2000	*1500	1125	113	3
at stage 2 to 3	2000	1000	*875	88	2.3
at stage 3 to 4	2000	1000	750	*413	11
at stage 4 to 5	2000	1000	750	75	*39

* The asterisk shows the first number in the horizontal line which is changed as a result of altering the mortality pattern.

So the answers to the questions are:

(a) Between stages 1 and 2 (1000 individuals):

(b) Between stages 4 and 5 (97 per cent):

(c) Halve the mortality between stages 4 and 5 (make it 48.5 per cent instead of 97 per cent).

SAQ answers and comments

SAQ 1 A defines a generation histogram (iv), e.g. Figure 1.1B.

B defines a total population curve (i), e.g. Figure 1.1A, the top line.

C defines partial population curves (ii), e.g. Figure 1.1A, most of the curves.

D defines a cumulative population curve (iii), e.g. Figure 1.1A, the heavy dashed curves for cumulative adult populations.

E defines a generation curve (v), e.g. Figure 1.1C.

If you had any difficulty with these, read *IPE*, Sections 1.3 and 1.4 again.

SAQ 2 (a) and (d) are TRUE.

(b), (c) and (e) are FALSE.

(b) Study Table 1.1 if you need to convince yourself that successive percentage mortalities can add up to much more than 100 per cent; their sum has no meaning.

(c) If log F is greater than K, then the population must increase and cannot be stable.

(e) On an arithmetic scale, changes at the lower end (e.g. less than 10) are scarcely noticeable if the scale goes up to hundreds; a logarithmic scale on the ordinate must be used to make changes in the low numbers as obvious as changes of the same proportion in the high numbers.

If you had any difficulty with these, read *IPE*, Sections 1.5 and 1.6 again.

SAQ 3 A defines intrinsic rate of natural increase (iii); see the beginning of Section 2.3.

B defines inverse density dependent mortality (vii), e.g. Figure 2.6B.

C defines density dependent mortality (i), e.g. Figure 2.6A.

D defines density independent mortality (ii), e.g. Figure 2.6C or D.

E defines scramble competition (v), e.g. Figure 2.11 curve B.

F defines contest competition (vi), e.g. Figure 2.11 curve A.

G defines a regulated population (iv), see Section 2.4.

If you had any difficulty with these, re-read *IPE*, Sections 2.3, 2.4 and 2.6.

SAQ 4 (b) is FALSE; there are conditions such as those given in (e), when the two will coexist.

The other statements are TRUE.

Competition is defined in *IPE*, Section 2.2 (the words in (a) are a modification of Bakker's statement). Section 3.9 includes statements (d) and (e), and statement (c) follows from these.

SAQ 5 This is really an exercise based on *IPE*, Figure 3.4, and the last two paragraphs of Section 3.4. Recall that when $\frac{dN}{dt} = 0$, the population of N cannot increase.

(a) (ii) Q will eliminate P as in Figure 3.4B if P is N_1 and Q is N_2.

(b) (iii) The species will coexist in a stable way as in Figure 3.4D.

(c) (iv) Either one of the species will eventually eliminate the other as in Figure 3.4C, but they may coexist for some time.

SAQ 6 Thompson's model includes statements (a), (c) and (e).

Nicholson's models include statements (a), (b), (c), (d) and (g).

Hassell and Varley's model includes statements (a), (c), (f), (g) and (h).

Statement (i) is probably true but not so far incorporated into a model.

Statement (h) defines Q, the Quest constant and statement (f) defines m, the mutual interference constant. Statement (d) defines Nicholson's constant a, the area of discovery. Statement (b) was coined by Varley to describe the effect of a Nicholsonian parasite; it should perhaps *not* be included in the list above.

SAQ 7 (c) is FALSE; the calculations associated with *IPE*, Figure 5.8, show no regulation of population by the random variable representing weather.

The other statements are all TRUE; weather can determine population fluctuations, but the regulation of the population about a stable level must be due to a density dependent mortality factor.

SAQ 8 Attributes of insects which make these suitable as possible agents for biological control include:

Specificity—the parasite or predator should confine its attention to the pest species.

High search efficiency—this would allow the parasite to reduce the host population to a very low level.

Mutual interference constant m within the range 0 to 1—this should promote stability in population level; a higher constant will lead to more effective spread of the parasite.

Tendency to aggregate—this leads to greater stability of populations.

Synchronization of life history with that of the host (unless the host is always available)—the story of control of *Promecotheca* (see *IPE*, Section 9.4.3) illustrates the importance of this.

Ability to withstand the climate into which it is introduced—the importance of this is illustrated by the story of *Aphytis lingnanensis* in California (see *IPE*, Section 9.4.1).

SAQ 9

List of aspects of models derived from laboratory studies	*List of aspects of models derived from field studies*
logistic curve of population increase	catastrophic effects of weather
intraspecific competition	area of discovery of parasites
Lotka-Volterra relationships including competitive exclusion and conditions for coexistence of competing species	life tables and how they vary with time and place
behaviour of parasites including their methods of searching and their interactions with other parasites	evidence for cyclic changes in numbers
effects of environmental factors on insect physiology and behaviour	evidence for success or otherwise of biological control

Units 8 and 9

Unit 8 Populations of plants and vertebrates
Unit 9 General principles of population regulation

Contents

Study guide to Units 8 and 9

These two Units should be studied together; they form an account and discussion of aspects of population regulation in flowering plants, fishes, birds and mammals.

Unit 8 is presented in the form of case studies of selected populations for which there is a reasonable amount of relevant information; in Unit 9 the processes of population regulation for the four groups of organisms are discussed, relying on the basic information given in Unit 8, with a few further examples. There are thus two possible strategies for working through these two Units: (1) you could read through Unit 8 fairly quickly and then more slowly through Unit 9, referring back to the relevant parts of Unit 8 when they are mentioned in Unit 9; (2) you could read through the Introduction and Section 8.1 of Unit 8 and then turn to Unit 9 and read through the Introduction (Section 9.0), then turn back to Unit 8 and read Sections 8.2 (on populations of terrestrial plants) and 8.2.1 (on buttercups) and then read Section 9.1 (on population regulation of higher plants)—and so on for fishes (Sections 8.3, 8.3.1, 8.3.2 and 9.2), birds and mammals. The first strategy is probably better if you can concentrate your study of the two Units into one week, but the second strategy might be more suitable if you must spread your study over three weeks (two for this Course and one for another) with two or three hours to spend at widely spaced intervals. The two Units together are designed to form less than two weeks' 'work', to give you time also to study parts of *ABE*.*

Section 8.0 introduces the broad categories of population studies. You have already studied life tables (in *IPE*) but the examples used there are insects with discrete annual generations; in Section 8.1 the use of life tables for organisms with longer lives and overlapping generations is discussed with examples of mammals and birds.

Section 8.2 concerns flowering plants with three species of buttercups growing in North Wales taken as a case study of how population numbers changed with time.

Section 8.3 concerns populations of fishes; the three species chosen as case studies are all British freshwater fishes: two carnivores feeding mainly on benthic invertebrates (perch and trout) and one piscivore (pike). The perch and pike populations have both been deliberately heavily fished for known periods of time and their responses to this are compared. Trout are an example of fish with territorial behaviour.

Section 8.4 deals with three species of British birds with contrasted feeding habits: the herbivorous grouse; omnivorous (but mainly insectivorous) great tits; and predatory, nocturnal tawny owls. The studies described have all been carried out for long periods of time and illustrate the importance of observing how the animals actually behave in their normal habitats.

None of the mammal populations studied in Section 8.5 are British. The small, cryptic voles present a contrast to the two large herbivores, American moose and African wildebeest. Lions and wolves are both 'top predators' living in social groups, but there are many contrasts between them. Studies of carnivores like these and their prey is very important when planning conservation programmes such as those for endangered species of large mammals.

Section 9.0 (the Introduction to Unit 9) outlines three contrasted views about population regulation of animals; you should note the differences and similarities between these.

A possible 'model' for comparing regulation of populations of flowering plants is discussed in Section 9.1. Similar 'models', showing stages at which population numbers may be regulated or determined, are presented in Sections 9.2 (for fishes), 9.3 (for birds) and 9.4 (for mammals). Each of these Sections refers you back to Sections of Unit 8 so that you can use the information and figures given there as part of the argument about mechanisms of population regulation.

Sections 9.5, 9.6 and 9.7 consider general aspects of population numbers. Maximum and actual rates of increase are discussed briefly in Section 9.5.

* The Open University (1974) S321/S323 *The Analysis of Biological Experiments*, which is part of the supplementary material for this Course.

2

Sections 9.6 and 9.7 are concerned with the two categories of consumers—herbivores and carnivores—and their food organisms; the discussion includes insect populations (treated in Units 6 and 7) as well as the vertebrate animals which are the main subjects of Units 8 and 9.

Section 9.8 summarizes the principles discussed in the two Units.

If you are short of time, you should omit the Sections concerned with plants (Sections 8.2 and 9.1); if very short of time, you should also omit the Sections concerned with fishes (Sections 8.3 and 9.2). You should study the Sections on birds (Sections 8.4 and 9.3) very carefully and read all the general Sections.

Note that for these two Units there are a common set of Objectives and Tables A1 and A2 given at the beginning of Unit 8. The references for both Units, and the SAQs and their answers are given at the end of Unit 9. The ITQs are numbered consecutively through the two Units and the answers are given at the end of Unit 9.

Objectives

After studying Units 8 and 9, students should be able to:

1 Define correctly and distinguish between true and false statements concerning each of the terms and principles listed in Table A1.
(Tested in SAQ 1.)

2 Given appropriate information:
(a) Draw up and analyse life tables and calculate or compare k-values and reproductive rates.
(Tested in ITQs and SAQs 1, 2, 4 and 5.)
(b) Identify key factors, regulating factors, density dependent factors, density independent factors and inverse density dependent factors.
(Tested in ITQs and SAQ 5.)
(c) Evaluate possible ways of altering the level of a population.
(Tested in ITQs and SAQ 5.)

3 Evaluate, by identifying the underlying assumptions, the methods used for sampling populations and assessing the age distribution of individuals in samples.
(Tested in ITQs and SAQ 2.)

4 Assess by compiling lists, selecting from a multiple choice situation or by writing essays:
(a) The part played by epideictic displays or territorial behaviour in processes of population regulation or change of fishes, birds and mammals.
(Tested in ITQs.)
(b) The relationships between population densities of herbivores and their food-plants.
(Tested in ITQs.)
(c) The relationships between population densities of carnivores and their prey species.
(Tested in ITQs.)
(d) The similarities and the differences between the population dynamics of named species of insects, fishes, birds and mammals.
(Tested in ITQs and SAQ 4.)

5 Apply the principles developed in this Block of Units to make hypotheses or design investigations or draw conclusions from data relating to situations or to species not treated in the Units.
(Tested in SAQs 4 and 5.)

6 Assess published data and statements relating to population dynamics in the light of principles developed in this Block of Units distinguishing between valid and invalid arguments and conclusions.
(Tested in ITQs and SAQ 3.)

Table A1

List of scientific terms, concepts and principles in Units 8 and 9

Introduced in Units 6 and 7 and *IPE*	Developed in Units 8 and 9	Page No.
intrinsic rate of increase	survivorship curves	5
k-values and key factor analysis	methods of sampling populations of plants, fishes, birds and mammals	10, 16, 22, 31
life tables	methods of deducing ages of fishes and mammals	6, 16, 49
scramble and contest competition	propagation of plants by seeds and vegetatively	10, 12
Verhulst–Pearl equation and logistic curves	territorial behaviour of vertebrate animals	21, 23, 27, 28, 33, 55
density dependence	epideictic displays of vertebrate animals	40
density independence	herbivore/plant relationships	53
inverse density dependence	carnivore/prey relationships	54
regulated population	population dynamics of:	
determination of population change	flowering plants	10, 41
use of population models to predict	fishes	15, 44
population levels and changes	birds	22, 46
	mammals	30, 49

Table A2

Units and Sections of Open University Science Courses taken as prerequisites

Course	Unit	Section	Objectives
S100	20	20.3	6
		20.4	7, 8, 9
		20.5	10
		20.6	12
		Appendix 1	6, 7, 8, 9
	21	21.3	2, 10
S2–3	Block 3	Part 2	2, 4, 5

8.0 Introduction

Studies of the changes in numbers of individuals in populations started comparatively recently. They fall into three broad categories:

1 Studies of the age of death of individuals in a particular area/population.
2 Studies of a population of which identified individuals are observed at frequent intervals throughout their lives.
3 Studies of the age distribution within samples of the population leading to estimates of shrinkage in numbers as the population ages.

Much of the work on insect populations has been of the third type, with samples taken from a population and examined carefully. From these samples the total numbers surviving at each time or stage are estimated for the whole population, and the detailed examination of the sample often indicates causes of mortality, e.g. the presence of parasites that would have killed the individuals later.

Only in laboratory situations is it possible to observe *all* the individuals in a population and, in some cases, to be able to identify the fate of each. Your Home Experiments on *Daphnia* and *Drosophila* show that even in cultures it may be impractical to follow the fate of individuals—they may not be easily recognizable and it may not be possible to make sufficiently frequent observations. How much more difficult this type of work is in field conditions!

Since few studies cover the fate of all individuals, there is immediately a 'sampling' problem—how representative are the observations of the situation in the population as a whole? How 'good' are the observations, have they a particular bias or can they be considered random? What sort of probability limits should be set on estimates of changes within a whole population from studies on part of it? You should consult *ABE* for a discussion of this type of problem; Part 3, Sections 3.4 and Part 4 are particularly relevant.

sampling problem

Another real problem in studies of a wild population is to know whether the period during which it was studied is 'typical'; there may be climatic factors that vary over short or long periods or there may be variations in populations of other organisms that directly or indirectly affect the ones studied. There are many other possible variables, and one problem to be faced before the study is to decide which variables to record and which to ignore.

8.1 Life tables and survivorship curves

Study comment You have already studied life tables of insects with discrete annual generations. This Section introduces life tables as presented in the literature for vertebrate animals, which typically have overlapping generations and breed more than once. You should note the different ways of obtaining life tables and the different types of survivorship curve; try to understand the biological significance of the differences between these.

Life tables summarize certain vital statistics of populations; they show for a group of individuals, called a cohort, the number of deaths for each interval of age, the rate of mortality, the numbers surviving and their further expectation of life. The cohort may be a real set of individuals or a composite group, depending on how the population has been studied. It is a 'real' cohort if the individuals represent a random sample of the population, all born in the same year; their ages at death are recorded. 'Imaginary' cohorts can be constructed by recording the age at death of a random sample of the population (not all the same generation or 'year class'), or by observing identified individuals of known ages in the population at frequent intervals through their lives. If these individuals can be put into definite age groups whose members are accounted for separately, then the data can be used to construct several 'real' cohorts provided that the study covers a sufficient length of time. It may be necessary to combine all the data to get a satisfactory life table, in which case this will cover an 'imaginary' cohort.

real and imaginary cohorts

The life table for a 'real' cohort is sometimes called a 'horizontal' life table, in contrast to the 'vertical' life table constructed by taking a census of a population of mixed ages and working out for each age interval the proportion of the individuals living at that time that die before reaching the next age interval.

horizontal and vertical life tables

TABLE 1 Life tables for male and female buffaloes from Rwanda, Central Africa. (From Spinage, 1972.)

x (years)	males d_x	males l_x	males q_x %	males e_x (years)	females d_x	females l_x	females q_x	females e_x
0–1	(70)*	1 000	50	6.0	(55)*	1 000	50	4.0
1–2	3	500	4.3	10.0	4	500	7.2	6.0
2–3	3	469	4.5	9.9	4	465	7.8	5.5
3–4	0	458	0	9.0	4	428	8.5	5.0
4–5	0	458	0	8.0	4	392	9.3	4.3
5–6	3	458	4.7	7.0	11	355	28.5	3.7
6–7	1	435	1.1	6.3	4	255	14.4	4.0
7–8	6	430	10	5.4	4	219	16.6	3.6
8–9	3	386	5.5	4.9	4	182	20	3.2
9–10	6	364	11.8	4.2	3	145	20	2.9
10–11	8	321	17.8	3.7	3	118	25	2.4
11–12	10	264	27	3.4	5	91	46	2.0
12–13	7	193	26	3.4	2	46	42.7	2.5
13–14	5	143	25	3.4	1	27	50	2.8
14–15	2	107	13.4	2.6	0	18	0	3.0
15–16	3	93	23	2.9	1	18	50	2.0
16–17	4	71	40	2.6	0	9	0	2.5
17–18	1	43	16.7	3.0	0	9	0	1.5
18–19	1	36	20	2.5	1	9	100	0.5
19–20	1	29	25	2.0		0		
20–21	1	21	33	1.5				
21–22	1	14	50	1.0				
22–23	1	7	100	0.5				

buffalo

Totals 70 + (70)* males   55 + (55)* females
Mean length of life 6.3 years   4.27 years

* These are estimates based on the 'reasonable assumption' that 50 per cent of calves die in the first year of life. This life table is based on examination of buffalo skulls found in the Akagera National Park.

ITQ 1 Suggest why vertical and horizontal life tables for the same species in the same area may be different.

Read the answer to ITQ 1 (p. 62).

Life tables are usually given as a series of columns as in Table 1. The cohort is usually assumed to start with 1 000 (sometimes 100) members. The symbols mean:

life table symbols

d_x number of deaths at age x (often these figures are the 'raw' data of the table)

l_x number of survivors from the cohort of 1 000 at the beginning of age interval x

q_x mortality rate during interval x; this is the fraction of those living at the beginning of the interval which die during the interval.

e_x expectation of life of those individuals surviving to the beginning of age interval x; this is the mean life span remaining to those attaining age x. Calculation of e_x is quite complicated except by the graphical method of measuring the area under a survivorship curve beyond age x and dividing this by the value of l_x.

ITQ 2 Plot the survivorship curves for male and female buffaloes (using the data in Table 1) on the semilogarithmic graph paper in Figure 1. Describe the pattern of mortality in the two sexes.

Read the answer to ITQ 2 and compare your curves with the completed Figure 1 on p. 17.

Spinage's data for buffaloes show marked differences in survival pattern between males and females, and changes in rate of mortality as the cohort ages. Field investigations might reveal possible reasons for the change in slope of the survivorship curve at certain ages, e.g. possibly female buffaloes start to breed

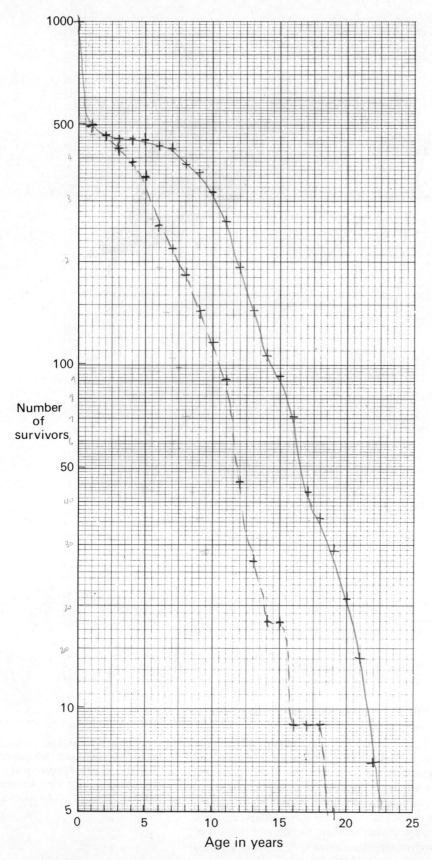

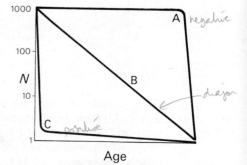

Figure 2 Three types of survivorship curves: ordinate—N (number of survivors); abscissa—age.

at five years old and males at nine years old (but it seems more probable that both sexes begin to breed when younger than this). Possibly their group behaviour alters as they grow older, and is different for males and females.

Deevey (1947), in a thoughtful review of life tables, pointed out that there are three theoretical types of survivorship curve; these are shown in Figure 2.

ITQ 3 Describe in words the pattern of mortality which is illustrated by each of the curves A, B, and C.

Read the answer to ITQ 3.

7

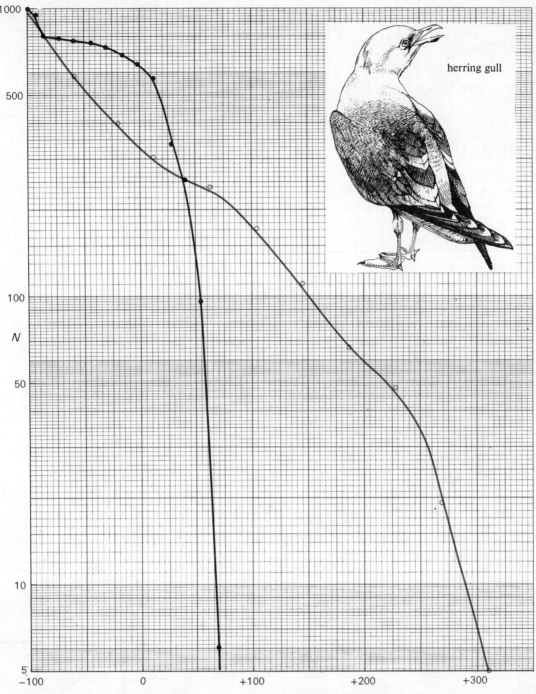

1000

500

N

100

50

10

5

−100   0   +100   +200   +300

Percentage deviation from mean length of life

herring gull

Few life tables fit exactly any of these three patterns. The buffalo female curve (see p. 17) could be interpreted as a diagonal type by drawing a straight line between 1 000 at age 0 and 5 at age 20. The male buffalo curve shows a positively skew rectangular shape up to age 9 and then a negatively skew rectangular shape. Where the animals' life history includes a vulnerable larval stage (e.g. the planktonic larvae of barnacles or mussels or plaice) the survivorship curve will be positively skew. Often no data are available for production and survival of very early stages, and then the survivorship curve (based only on older animals) may appear to be diagonal or even negatively skew rectangular.

Life spans of animals vary from less than a day (for tiny protistans) to a century or more. To compare survivorship curves for such different periods of time, it is convenient either to take the mean length of life of the cohort as zero and to express each age as its percentage deviation from the mean, or to take the life span as 100 per cent and express each age as a percentage of the total life span.

These two methods of expressing survivorship are shown in Figures 3 and 4, using data for the herring gull *Larus argentatus* and for the Dall mountain sheep

Figure 3 Survivorship curves for herring gulls (red) and Dall mountain sheep (black) with age expressed as percentage deviation from the mean length of life of the cohort. (From Deevey, 1947.)

life span

8

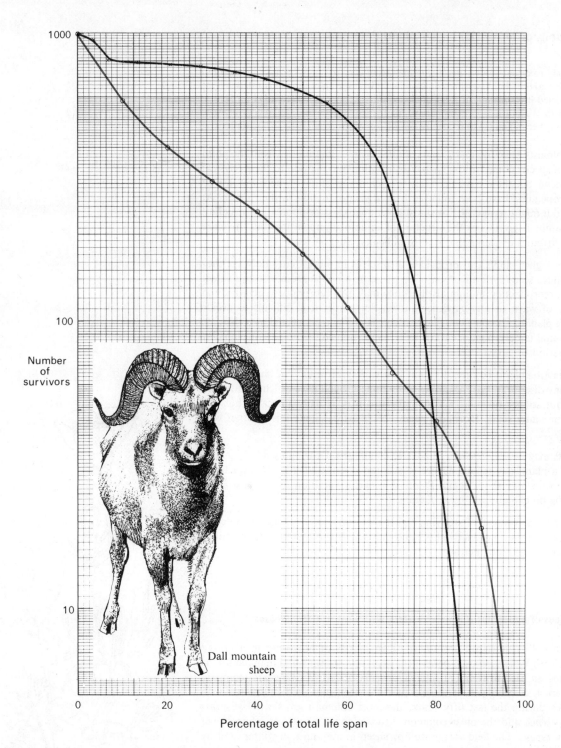

Number
of
survivors

Dall mountain
sheep

Percentage of total life span

Ovis dalli. The latter is a Bighorn sheep found in the Rocky Mountains and the data used here came from a study of skulls collected in a National Park in Alaska; the sheep were aged by comparing the size of the horns and state of the teeth. Look at the differences in the shapes of the curves using the same sets of data.

Figure 4 Survivorship curves for herring gulls (red) and Dall mountair sheep (black) with age expressed as percentage of the total life span of th cohort.

ITQ 4 Both life tables are based on relatively small numbers (608 sheep skulls and 1 252 dead gulls)—if further studies were made, would you expect survivorship curves based on larger numbers of individuals to look very different from those in Figures 3 and 4?

Read the answer to ITQ 4.

The data plotted against age as a percentage of the total life span emphasizes the difference between the negatively skewed rectangular type curve for the sheep and the diagonal type curve for the gull, but it does not make the point that some gulls have much longer than average lives (three times as long) whereas few sheep live for longer than 50 per cent more than the average life span.

Now you could attempt SAQs 1 and 2 (p. 61).

9

8.2 Populations of terrestrial higher plants (Angiosperms)

Study comment Flowering plants may seem easy to study because they do not move about, but they present certain difficulties; these are illustrated by the detailed investigation of populations of three species of buttercups in a field in North Wales (Section 8.2.1). Look for the similarities and differences between these species and try to recognize the factors which regulate their populations.

Plant ecologists have for many years studied plant communities from the point of view of species composition, spatial distribution and relation to their physical environment, but there have been few 'demographic' studies where the fate of plant 'propagules' have been followed from their production to their death. At first sight, it might seem easier to study a plant age class or population structure than an animal age class or the structure of an animal population, but there are certain attributes of plants that lead to difficulty. Suggest some!

1   Many higher plants can propagate themselves in more than one way; they may produce seed (by fertilization of an ovule by a pollen grain, the equivalent of sexual reproduction in animals) but they may also reproduce vegetatively in a variety of ways, as any gardener knows. Sometimes it is difficult to decide when one plant has become two plants; they may remain connected by a rhizome (a horizontal underground stem) or by a lateral branch of the parent plant for a long period of time.

2   Plants are able to become dormant in a variety of ways. The shoot with its leaves may disappear, but there may be an underground part which is alive and will sprout sooner or later. Seeds in the soil may remain dormant, but still viable, for many years. Dormant stages should be counted; this is often very difficult.

3   Plants may be very resilient in adverse conditions. Grazing by a large herbivore on a plant may reduce its size to such an extent that it is inconspicuous, but this plant may shoot again. Plants show great variation in size and form, depending on climatic and other factors.

Yet the great advantage of studying higher plants is that they remain rooted in one place over their productive lives and individuals can therefore be identified and observed over long periods.

8.2.1   Populations of buttercups *Ranunculus acris, R. repens* and *R. bulbosus*

Sarukhán (1971) described a study of three species of buttercup in a grazed field in North Wales. He used permanent sites for quadrats* and mapped all the buttercups on each at frequent intervals over a period of two years (compare *ABE*, Part 4, Section 4.2). The field had not been ploughed or treated with herbicides during the last fifty years; the most common grass was rye grass *Lolium perenne* and the most common dicotyledonous plant was white clover *Trifolium repens*. The field was grazed by sheep in autumn and winter, and by cattle and some horses during the summer. The three buttercup species are: *Ranunculus acris, R. repens* and *R. bulbosus* (which actually has a corm, not a bulb). All tend to form clumps of plants of the same species. Seeds of *R. repens* (creeping buttercup) germinate only when there is sufficient water in the soil and the plants are able to grow in waterlogged areas; seedlings of *R. bulbosus* (bulbous buttercup) grow only in well-drained areas, whereas seedlings of *R. acris* (meadow buttercup) have intermediate requirements.

Table 2 gives figures for three of the plots studied; in each of these, one species was present in moderate density. Note that these are field observations and that there was variation between plots.

ITQ 5   Fill in the spaces in the rows (c), (d), (h) and (k).

Read the answer to ITQ 5.

* A quadrat is a sample square (usually with sides 0.5 or 1.0 m long).

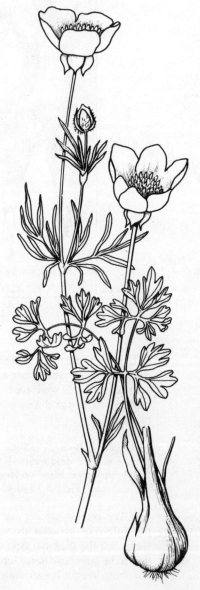

bulbous buttercup

10

TABLE 2 Data for three species of buttercup in a field in North Wales. (Each set taken for a plot where that species had a 'moderate' density.)

		Ranunculus species:		
		acris	*bulbosus*	*repens*
(a)	Number of plants at start of observations m⁻²	62	90	117
(b)	Number of plants at end of observations (2 years later) m⁻²	124	64	139
(c)	Net change (b−a)	62	−26	+12
(d)	Rate of increase (b/a)	2	·7	1·2
(e)	Number of plants that arrived during the two years (by all methods)	457	84	244
(f)	Number of plants lost during the two years	395	110	222
(g)	Number of plants present at the beginning and still alive at end	34	41	13
(h)	Percentage survival of plants present at beginning (100g/a)	54·8	455	11·1
(i)	Expected time for a complete turnover of plants (200/(100−h)) years	4.3	3.7	2.25
(j)	Total plants recorded during the two years	519	174	361
(k)	Percentage annual mortality of all individuals (100f/j)	76·1	63·2	61·5

TABLE 3 Survivorship data for cohorts of buttercups on the plots used for Table 2. The study began in April 1969; the figures in columns headed *T* are weeks after the start of the study.

Ranunculus acris			*R. bulbosus*		*R. repens*	
T	mature plants	seedlings emerged at *T* = 0	*T*	all plants	*T*	all plants
0	62	126	0	90	0	117
3	56	76	4	88	3	116
6	52	60	5	86	5	111
9	51	52	8	81	9	109
11	48	38	10	76	12	102
13	42	27	13	75	14	98
15	40	25	15	70	17	97
18	39	23	32	70	28	77
33	38	19	42	69	30	70
43	38	18	50	69	32	56
52	36	17	54	67	55	27
56	36	15	58	64	61	25
58	36	14	59	61	65	23
62	33	14	62	55	67	23
68	33	14	66	52	69	19
75	30	14	77	50	75	15
100	28	12	81	50	78	13
107	22	12	107	41	106	13
					118	10
					121	8
					123	0

The census figures were used to produce survivorship tables as illustrated in Table 3.

ITQ 6 From this Table, try to make statements about the pattern of mortality of the three species of buttercups. You may find it helpful to plot survivorship curves.

Read the answer to ITQ 6.

meadow buttercup

11

The buttercup plants were growing at different densities in different plots. Figure 5 shows the relationship between original density and the relative increase in numbers of plants over the two years for *R. repens*.

ITQ 7 What conclusion would you draw from this Figure?

Read the answer to ITQ 7.

The percentage survivorship figures for all these *R. repens* populations were fairly similar, between 6 and 13 per cent over the two years, and the maximum duration of life was under three years; percentage annual mortality was between 53 and 79, and more than half the original plants had been replaced by the second year. In this species, vegetative reproduction contributes much more to the population than does seed production; most of the mature plants were replaced by 'daughters' shortly before they died. The typical age distribution just before the growing season was: 70 per cent of the population were in their first year and had been formed vegetatively; 20 per cent were in their second

vegetative reproduction

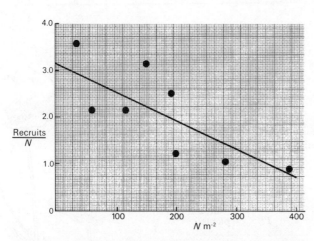

Figure 5 The number of plants of *Ranunculus repens* recruited to the population expressed as a proportion of the original plant density (Recruits/N m^{-2}) plotted against the original plant density (N m^{-2}). (From Sarukhan, 1971.)

Figure 6 The life expectancy (in weeks) of vegetative daughters of *Ranunculus repens* plotted against the average density of plants in April of 1969, 1970 and 1971, expressed as N m^{-2}. (From Sarukhan, 1971.)

year; and about 5 per cent were in their third year; the remaining 5 per cent were new seedlings. The seedlings have a low expectation of life; only one seedling from all the sites attained more than 2 years of age. Figure 6 shows the life expectancy of vegetative propagules (daughter plants) of *R. repens* plotted against the density of plants at that site. The graph shows an inverse relationship; the more crowded plants have a shorter expectation of life, i.e. mortality is density dependent

Ranunculus acris and *R. bulbosus* present a contrast to *R. repens* because, for both, reproduction by seeds produces more new individuals than vegetative reproduction (which does not occur at all in *R. bulbosus*). The numbers of one-seeded fruits (achenes) produced by all the plants in the quadrats were noted, but investigation of the seeds in the soil was carried out in other parts of the field because it involved disturbance. Seed collected from plants in the field were sown in small 'subquadrat' areas; flower heads were removed from all plants close to these areas to ensure that emerged seedlings would probably have been derived from the planted seeds. Seedlings were counted 15 days later; soil samples were taken and carefully sieved to recover all buttercup seeds, which were then planted in pots and the numbers of seedlings were recorded over the following four weeks. Again any seeds in the soil were isolated and examined and tested chemically for viability. Four categories of seeds were thus recognized: the viable seeds that germinated in the field; viable dormant seeds that germinated when sown in the pots; dormant seeds that did not germinate in the pots but were still viable and could have germinated later; dead seeds. In another set of observations, samples of soil were examined for buried seed populations occurring naturally, and these were categorized as dormant seeds

viable and dormant seeds

and dead ones. One unexpected problem arose in the area where seeds were sown;
it was protected from birds by wire netting which also prevented grazing by
cattle. The grass grew taller and voles were able to penetrate the area and ate
some of the achenes; they sliced them neatly and removed the contents.

Some of the observations on flowering and setting seed are given in Table 4.

TABLE 4 Data about flower and achene (seed) production of buttercups in the plots
in North Wales.

	Year	Number of flowers m^{-2}	Number of achenes (seeds) m^{-2}
R. repens	1969	50.0	149.5
	1970	12.4	16.6
R. bulbosus	1969	111.3	982.5
	1970	69.2	162.9
R. acris	1969	110.0	800
	1970	42.2	15.6

Notice the enormous differences in flower and seed production between the two
years; probably this was due to differences in timing of heavy grazing in relation
to flowering of the plants. Weather factors may also have operated. The plants
of R. repens produced an average of 115 vegetative daughters per square metre
in 1969 and an average of 114 in 1970. The grazers eat off the flowering stalks
but do not seem to interfere with the production of daughter plants.

Some of the data about the buried seed population in the soil are given in
Table 5.

TABLE 5 The buried seed populations found in three different sites in March 1970
(each figure is the average of five soil cores).

Predominant species on the surface	Empty seeds	Living seeds m^{-2} R. acris	R. bulbosus	R. repens
R. acris	1 945	458	115	57
R. bulbosus	1 945	172	515	57
R. repens	115	115	57	572

ITQ 8 From the data in Table 5 what conclusions would you draw about the
distribution of seeds from their parent plants?

Read the answer to ITQ 8.

Comparison of Tables 4 and 5 reveals that the number of seeds of R. repens in
March 1970, in the soil of a plot dominated by that species, is higher than the
average production of seeds m^{-2} by R. repens in 1969. For the other two species,
the corresponding values for buried living seeds are just over half those for
average seed production. Many flower heads of these two species (which are
taller than R. repens) are eaten by cattle. If fully mature, seed can pass through
cattle guts and remain viable; they may be dispersed in this way, but are concentrated because voided as part of cow pats. Rodents may eat seeds on the
plants and so may some birds; some seeds will be destroyed and some are
dispersed by these agents.

In the soil, seeds of R. repens suffered the greatest predation; more than 50 per
cent were eaten in some plots and less than 10 per cent germinated. For R. acris,
the percentage germination was about 50 per cent, and less than one quarter of
the seeds disappeared, presumably by predation; for R. bulbosus, the situation
was intermediate between the other two species with more than 20 per cent
germination and about 30 per cent being eaten.

The seedlings of R. acris and R. bulbosus showed a very rapid decrease in
numbers for the first few weeks after germination, but then the mortality rate
became similar to that of the mature plants. Look back to your survivorship
curves based on Table 3 to find support for this statement.

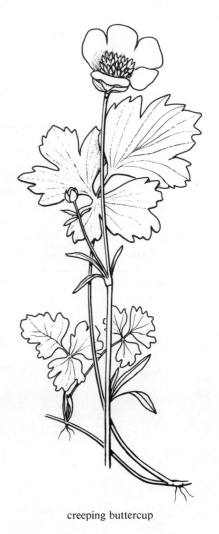

creeping buttercup

13

Sarukhán constructed population flux models from his data for the three species of buttercups; these are presented in modified form in Tables 6, 7 and 8. Note that none of these models give a stable population.

ITQ 9 From these Tables and the information given earlier:

(a) When is the greatest mortality of plants for each of the species?

(b) When is the greatest mortality of seeds for each of the three species?

(c) Compare the potential contribution of seeds and of 'daughters' for these three species.

Read the answer to ITQ 9.

TABLE 6 Population flux model for *Ranunculus acris*. The values are individuals m⁻² at the beginning of the month, or additions and losses during the interval since the last entry.

Month	Seeds in soil		Plants	
	died in interval	viable	live	died in interval
April		500	57	
			+	
		seedlings →	176	
	224			126
June		100	107	
		+		
		1000 ← seeds		
	10			26
August		1090	81	
	345			
October		745	72	9
			+	
	48		7	4
			daughters	
December		697	75	
	137			13
April		560	62	
	Total seeds died 764		Total plants died 178	

TABLE 7 Population flux model for *Ranunculus repens*. The values are individuals m⁻² at the beginning of the month or additions and losses during the interval since the last entry.

Month	Seeds in soil		Plants	
	died in interval	viable	live	died in interval
April		367 ── seedlings ──	154	
		+	+	
			→ 25	
		84 ← seeds		
	106			45
June		320	133	
	113			25
August		207	108	
	0			39
October		207	69	
			+	
	21		102	27
			daughters	
December		186	144	
	6			21
April		180	123	
	Total seeds died 246		Total plants died 158	

TABLE 8 Population flux model for *Ranunculus bulbosus*. The values are individuals m^{-2} at the beginning of the month or additions and losses during the interval since the last entry. Note that there is no vegetative reproduction by this species.

Month	Seeds in soil died in interval	Seeds in soil viable	Plants live	Plants died in interval
April		400	43	
	148 +	565 ← seeds		5
June		817	38	
	104			4
August		713	34	
	463			2
October		250	32	
		seedlings → 95 +		
	5			41
December		150	86	
	19			20
April		131	66	
	Total seeds died 739		Total plants died 72	

The three species of buttercup are closely related, but there are striking differences in the population flux models. The relative importance of seedlings and of 'daughters' (vegetative propagules) differs among the three, and so does the timing of flowering, seed production and seed germination. Study of the models indicates stages of the plants which would be worth further study in an attempt to determine the precise factors which lead to death of mature plants, seedlings and daughters.

8.3 Populations of fishes

Study comment Fish populations are of economic interest, but fishery statistics usually give data of limited value for interpreting mechanisms of regulation. The perch and pike populations in Windermere (Section 8.3.1) have both been heavily fished, but with different consequences; the two must be considered together because the pike prey on the perch. You should look for the similarities and differences between them and compare and contrast them with the brown trout (Section 8.3.2), a fish which exhibits a very different type of behaviour and survivorship pattern.

Many natural populations of fishes support economically important fisheries, so there has been interest in their sizes and fluctuations over many years. Catch statistics have been available for a long time, but need to be interpreted carefully because of changes in the numbers of fishermen and in the type of fishing gear in use, and consequently in the efficiency with which the fishermen 'sample' the population. From the early days of fishery science, there have been attempts to analyse the catch into age groups, and this has revealed that the relative numbers in different age groups often fluctuated very markedly over a period of time.

When biologists began to study commercial fish species, it soon became clear that very young individuals often lived in a completely different way from the older, catchable fish. Some marine fish, e.g. plaice, lay eggs that float and soon hatch into tiny larvae which live and feed in the plankton. When they reach a certain size (and age), they undergo metamorphosis and start to feed and live, like adults, on the bottom. Plaice *Pleuronectes platessa* illustrate metamorphosis of an extreme type, but many fishes show definite changes in shape and habit at a certain stage of their lives; they may undergo migrations, like the salmon *Salmo salar* and eel *Anguilla anguilla*. Fish may spawn annually for many years, growing in weight between each spawning cycle; the larger the fish, the greater the number of eggs that she can lay, so the potential recruitment to a population must take account of growth and longevity. If the fishery is to maintain itself,

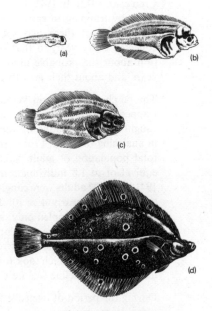

plaice metamorphosis
(not to same scale)

it is essential that a proportion of the population should spawn at least once and it is therefore sensible to set a lower limit to the size of fishes which may be landed and sold. Most fishes that live in temperate or polar waters show cyclical changes from rapid to slow growth related to seasonal changes. This cycle of changing growth rate is often recorded in hard parts of the body, such as bones, scales and otoliths (small calcareous concretions in the inner ears), and this makes it possible to deduce the age of a fish and may make it possible to calculate the size it had attained in earlier years of its life (see Unit 3, Figure 10). Thus, from a sample of fish, it is possible to obtain information about:

deducing age of fishes

(a) the actual size distribution in that sample;

(b) the age distribution of the fishes caught;

(c) the past growth history of those fishes;

(d) the state of maturity of the individuals;

(e) feeding habits and the presence of parasites.

The most difficult problem in studying fish populations is that of sampling: how closely does a sample collected by using a net or using an electric fishing machine conform to a random sample of the whole population of fishes? It is rarely possible to catch all the individuals or to follow them through their whole lives, but it is possible to mark individuals (there are many methods for doing this, each with snags) and to recapture these at intervals. 'Mark-recapture' methods form the most usual type of sampling from which population estimates are derived. You should refer to *ABE*, Part 4, Section 4.3 for a discussion of this type of sampling.

Electric fishing is shown in TV programme 4 of the Environment Course S2–3.

To study any species of fish, as for other organisms, it is necessary to know the phenology and natural history; in particular, it is necessary to know where and when the fish spawn and where they go to feed. Samples taken at spawning time give information about the potential recruitment that year and also about survival to spawning of the age groups represented in the sample of adult fish. Samples taken at other times may reveal that there is a heavy mortality at definite stages or in definite places. Since it is extremely difficult to sample the very young stages, there is often a gap in the estimates between the egg stage and the juvenile stage (young fish feeding in the same place and way as adults); sometimes there is a gap between the egg stage and the age of first breeding.

8.3.1   Populations of perch *Perca fluviatilis* and pike *Esox lucius* in Windermere

Windermere supports large populations of char *Salvelinus alpinus willughbii*, brown trout *Salmo trutta*, perch and pike; there is evidence that all these have been fished for food for a long time. The sport fishery became more important after about 1850, and commercial netting stopped completely in 1920 (see Le Cren *et al.* 1972). In 1941 the Freshwater Biological Association began to remove perch by trapping in early summer, and in 1944 they began to net pike. The populations of these two species have been followed by studying samples collected every year, so that there is now available a great deal of information about the probable numbers of fish of different ages present in different years and about their growth rates.

The perch are caught in Windermere by traps which are selective in that the trapped fish are between 9 and 29 cm long, but rarely longer. The samples caught in May and June are very largely males and this bias has to be recognized in analysing the data. From samples caught before 1941 it is estimated that the total population of adult perch (two years and older) was 5×10^6, which is equivalent to 1.8 individuals m^{-2} of shallow water (less than 10 m deep). In 1941, 1×10^6 adults were caught in the North Basin, and by 1947 over 90 tonnes (equivalent to nearly 4×10^6 fish) had been removed. Since 1948, the North Basin has been sampled only, and heavy fishing stopped in the South Basin in 1964. As expected, the population fell drastically between 1941 and 1944, but since then it has remained at round about 20 per cent of the former numbers, with fluctuations which took it up to 67 per cent of the 1941 population in 1959, followed by a decline (see Le Cren, 1955, 1958).

perch trapping

During the period of intensive fishing, the average size of adult perch fell as the older, larger members of the population were removed; the growth rate of the young adult perch showed an increase as the density of adults fell. In 1955 the

perch

16

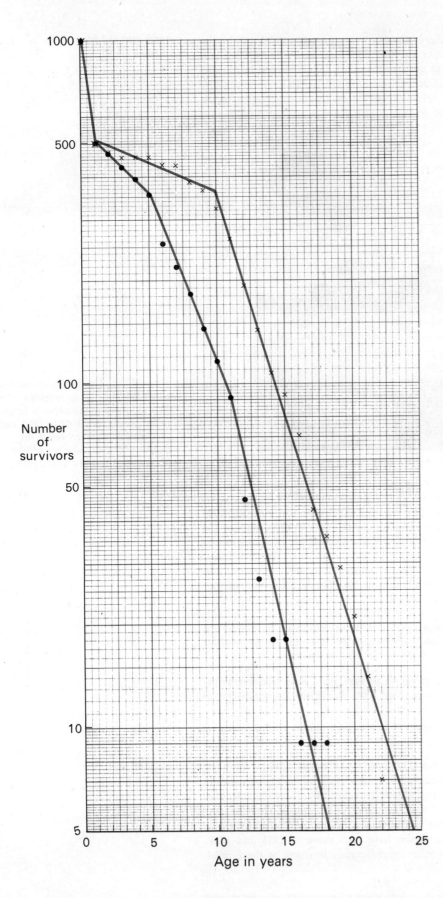

Number of survivors

Age in years

Figure 1 (completed) Survivorship curves for male (×) and female (●) buffaloes.

annual weight increment of three-year-old perch was six times that of perch of the same age in 1940 (Le Cren 1958); there was no apparent change in growth of perch in their first year of life and very little change in the second year.

ITQ 10 Suggest an explanation for this.

Read the answer to ITQ 10.

17

Older perch feed on benthos, the organisms living on the substratum in the shallow parts of the lake; they also eat some zooplankton. Perch eggs are laid in ribbons among weeds; the eggs are 2 to 2.5 mm in diameter and hatch into small fry which feed on very small zooplankton organisms. These fry eventually change their habit and become juveniles feeding on the bottom fauna or on larger zooplankton, i.e. on similar food to adults. Perch lay about 200 000 eggs per kg of body weight.

perch eggs and fry

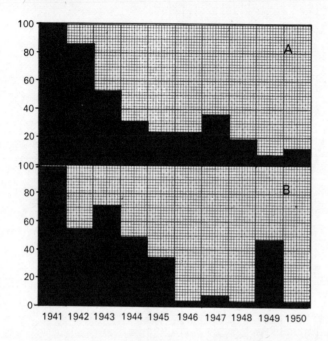

Figure 7 (A) The number of perch eggs produced and (B) the number of three-year-old perch in Windermere belonging to the year-classes 1941 to 1950, each expressed as a percentage of the numbers for the year-class of 1941. (Rearranged from Le Cren, 1955.)

There are no detailed figures from which to construct life tables for perch, but there are estimates of the total egg production in each year and then estimates of the population of young adults breeding three years later. Compare the histograms in Figure 7 A and B. These show how the number of eggs laid and the number of three-year-old perch varied from year to year. The histograms are referred to the year in which the eggs hatched, i.e. to their 'year-class', so that the number of three-year-old fish is shown immediately below the number of eggs from which they were derived.

ITQ 11 The number of three-year-old perch declined from a high level for the 'year class of 1941' (the fish hatched from eggs laid in 1941) to a low level for the year-class of 1945. Suggest an explanation for this decline.

ITQ 12 For the year-classes from 1945 to 1950, there were variations in the numbers of three-year-old perch surviving. Does the explanation you gave as an answer to ITQ 11 also explain these variations?

Read the answers to ITQs 11 and 12.

The perch population since 1945 has shown the phenomenon called 'dominant year-classes'; that is, the adult population in any year consists mainly of members of one particular year-class; fish hatched in other years contribute much smaller proportions to the adult population than the dominant year-class. The year-class of 1949 remained dominant in the adult perch population between 1952 and about 1958; other dominant year-classes have been those of 1955 and 1959. Dominant year-classes for perch occur synchronously in many of the English Lakes.

dominant year-classes

ITQ 13 How do you interpret this synchronous occurrence of large year-classes?

Read the answer to ITQ 13.

The pike are caught in gill-nets between October and February in shallow areas round the lake; the nets catch fish of 55 cm or longer (males of four years or older and females of three years or older). In 1944 to 1945 the catch was 756 pike weighing about 3 tonnes; since then the average catch has been about

pike netting

300 pike weighing about 1 tonne. Pike have been marked and released in spring and up to 80 per cent of these have been recaptured in the gill nets, suggesting that these are very efficient in catching the fish (see Kipling and Frost, 1970).

It is estimated that there were about 2300 pike aged two years or older in the lake in 1944, and the numbers of these fell until 1949, when they rose again above the 1944 level. This was the result of a large year-class (1947) coming into the catchable population. The years 1949, 1955 and 1959 also led to very large year-classes, and in 1961 the total number of pike was about twice that in 1944. The total biomass of these fish in the period 1961 to 1963 (associated with the dominant year-class of 1959) was almost the same as in 1944. The years when these large year-classes were produced all had high temperatures in the summer, when the young pike would be feeding on zooplankton. Note that 1949, 1955 and 1959 were also 'good' years for perch.

pike

Pike become piscivorous when they reach about 10 cm in length and eat perch, char and trout; the larger pike eat larger fish and the removal of old, really large pike from Windermere has probably allowed better survival of large trout and char.

The removal of perch and of pike from Windermere has been followed by quite different changes in population.

> **ITQ 14** Summarize the results of the removal operations on the populations of perch and of pike in Windermere.
>
> *Read the answer to ITQ 14.*

For both perch and pike there is evidence that the main fluctuations in populations are related to the success of the young fish, to the rate of mortality between the laying of the eggs and the coming of the fish into the sampled population. If life tables were available to be analysed, this would probably be the key factor. But what are the factors regulating the level about which the fluctuations occur?

fry survival

It seems likely that the population of pike in Windermere is now regulated by the Freshwater Biological Association; they net the pike heavily and kill a high proportion of the fish that reach 60 cm in length. This mortality is density dependent if the Association catch relatively more fish from an abundant population of pike over 60 cm long than from a sparse population. The removal of the larger fish probably leads to increased survival of smaller fish, and the numbers of pike in the lake are now higher than before the netting programme began. The many smaller pike generally equal about two-thirds of the biomass of the fewer larger fish of thirty years ago.

It has been suggested that the lower level about which the perch populations now fluctuate is the result of the change in pattern of pike populations.

The present pike population probably eats nearly the same weight of fish as that of thirty years ago, but the present pike are smaller and must eat smaller prey. Pike feed predominantly on perch during the summer; this is the season when the pike are growing fastest and taking the greatest quantity of food. The smaller, fast-growing pike probably take many more young perch than used to be taken by the relatively older and larger pike, and thus the perch population is regulated at a lower level; survival of the older, larger perch should be better than it used to be.

pike predation on perch

> **ITQ 15** Suggest what might be the changes in the trout and char populations since pike netting started in 1944.
>
> *Read the answer to ITQ 15.*

8.3.2 Populations of brown trout *Salmo trutta*

Allen (1951) made the first detailed study of population numbers in trout populations; he studied the fish living along a small stream in New Zealand, the Horokiwi. He estimated the population by the mark-recapture method, catching the fish by netting; he estimated the initial population by counting redds (places where the fish bury their eggs) and by computing the fecundity of the female fish. He went to great pains to estimate the probable errors of his values; these were usually of the order of ±50 per cent. Smoothed values from Allen's work are as follows:

New Zealand trout

Total egg production	900 000
Total fry ready to start feeding	500 000
Total fry 3 months after starting to feed	12 000
Total fry 6 months after starting to feed	7 500
Total fry one year after eggs are laid	4 500
Total number of two-year-olds	900
Total number of three-year-olds	180
Total number of four-year-olds	36
Total number of five-year-olds	7

Trout lay eggs which are comparatively large (about 5 mm in diameter); these are buried in gravel and hatch to give 'alevins' which have a large yolk sac. The alevins burrow more deeply into the gravel but, when the yolk is almost used up, they move towards the surface and emerge to poise themselves just above the gravel from where they take particles drifting in the water; these include invertebrates. The trout feed on similar food for the rest of their lives. They lay on average about 1 600 eggs per kg of body weight.

ITQ 16 From Allen's values given above, calculate the mortality between each stage of the life history of trout in the Horokiwi River.

Read the answer to ITQ 16.

Of the population at the start of the year-class, 99.5 per cent died before the end of the first year, but after that about 20 per cent of the individuals alive at the beginning of the year survived to the end of it. This is an example of a positively skew rectangular survival curve.

These figures for a New Zealand river can be compared with two sets of figures for British (Lake District) waters:

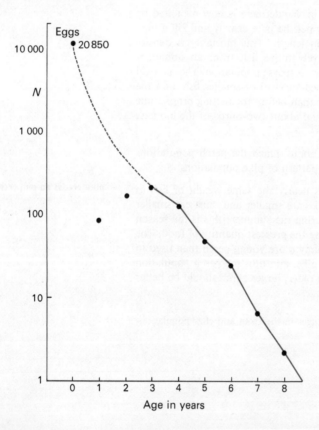

brown trout

Figure 8 A survivorship curve for brown trout in Three Dubs Tarn. (From Frost and Brown, 1970.)

(a) Three Dubs Tarn has an area of 16×10^3 m^2 and lies in moorland to the west of Windermere. The population was sampled annually by seine netting; other trout were caught by anglers at other times of the year. The same effort was put into the netting each year, and it is therefore assumed that the catch was proportional to the total population in that year. A life table was constructed

English tarn trout

20

by expressing the number of fish in one age group as a percentage of those in the preceding age group (Frost and Smyly, 1952; Frost and Brown, 1970). The number of survivors of different ages is plotted in Figure 8; note that the fish less than three years old are not sampled adequately by the seine net so that the early part of the curve is derived from an estimate of the egg production.

ITQ 17 Compare this survivorship curve with the figures for the Horokiwi River.

Read the answer to ITQ 17.

The variation among year-classes in Three Dubs Tarn suggests that the most successful (using the criterion of numbers reaching the size sampled by the netting) consisted of more than seven times as many fish as the least successful. These trout probably spend most of their lives in the Tarn; they spawn in small inflowing streams.

(b)   From combined observations on several small streams (becks) in the Lake District, Le Cren (1965) deduced the following average life table:

English stream trout

Number of eggs laid	800
Number of fry ready to start feeding	750
Number of yearlings	20
Number of two-year-old trout	10
Number of three-year-old trout	5
Number of four-year-old fish breeding	2

It is assumed that one fish is caught by anglers when three years old but otherwise the mortality is 'natural'.

ITQ 18 Compare the survival of the trout in these streams with that in the Horokiwi River.

Read the answer to ITQ 18.

Le Cren (1965) carried out some experiments on survival of fry in a small stream in the Lake District. He used grids to divide this stream into sections of equal length and stocked these with different numbers of eggs or fry. The results of a typical experiment are shown in Figure 9.

ITQ 19 Interpret Figure 9 from what you have read about competition.

Read the answer to ITQ 19.

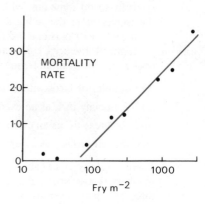

Figure 9 Instantaneous mortality rate (Z) of brown trout fry in a small stream plotted against the original density of the fry (N m^{-2}). (From Le Cren, 1965.)

Le Cren observed that most of the mortality happened between twenty and forty days after the fry began to feed; the dying fry had stomachs empty or nearly so and weighed less than the average starting weight, so it seems reasonable to assume that starvation was the main cause of the mortality. Kalleberg (1958) described how trout and salmon fry took up territories, within which they fed, and 'threatened' other fry which entered these areas. Fry that did not establish themselves in territories were not able to collect enough food for survival and moved off downstream. The size of the territories depended on how far the fry could see, so that there were more smaller territories in an area where large boulders broke up the 'landscape' than in an area where view was restricted only by the physical properties of the water. Kalleberg also showed that water movement affected territory size and that fry in still waters did not have the territorial habit.

territorial behaviour

It appears then that there is an upper limit, set by physical features of the environment, to the number of young trout that can establish themselves in a stream. If egg production in that stream is very high, there will be a very high mortality of young fry which fail to obtain territories; if egg production is comparatively low, the mortality of young fry should be very low since most may obtain territories. Contest competition at the stage when the fry begin to feed regulates the size of the population and continues to do so, since trout in rivers remain territoral except when seeking spawning areas. The territories become larger as the trout grow, but an old trout will tolerate trout of younger age-classes within its territory. Mortality of older trout may be related to catastrophes such as floods, or to predation or to a lack of food. Note that territory size is apparently not directly related to the amount of food present.

Now you could attempt SAQ 4.

8.4 Populations of birds

Study comment The best documented examples of change and stability in vertebrate populations over periods of many years are probably for birds. You should, therefore, work through this Section carefully. The tawny owl study should be familiar to you from S100, Unit 20 (Appendix 1) and you should have that Unit beside you as you read Section 8.4.1. Great tit numbers vary greatly from year to year, and key factor analysis (described in Section 8.4.2) indicates some of the underlying mechanisms. Red grouse (described in Section 8.4.3) are herbivores cropping a small proportion of their food plant, yet the fluctuations in grouse numbers can be explained in terms of food supply and the behaviour patterns of the cocks.

The technique of 'ringing' birds provides a method for identifying individuals; there are various methods of trapping, suitable for different habits and habitats, and the birds can usually be released unharmed and, apparently, unaffected by their experience. Birds that are active by day may be easy to observe visually or can be located (and recognized sometimes) by their song; nocturnal birds and those that skulk in thick cover present special problems.

All birds lay eggs which must be incubated for a certain period. The young typically spend some time with their parents, either being fed by them or feeding themselves; later the young disperse. Adults may pair for life or they may pair for the whole or part of a breeding season. To follow changes in numbers of bird populations, the eggs, chicks and adults, paired or single, can be counted either once a year or at more frequent intervals.

As examples of birds with different types of habits we shall take the following:

(a) the tawny owl, as an example of a predatory bird;

(b) the great tit, as an example of an insectivorous bird typical of woodlands;

(c) the red grouse, as an example of a herbivorous ground-living bird.

There are many other habits among birds; detailed information about population numbers is available for a few of them, but here we have space only for the three examples given above.

8.4.1 Populations of the tawny owl *Strix aluco*

The most detailed study of any predatory bird, indeed of any predatory vertebrate, is that of H. N. Southern and collaborators in Wytham Wood, near Oxford. Detailed information from this study is treated as a 'structured exercise' in Appendix 1 (Black) of S100, Unit 20. *You are strongly advised to turn to that Unit and work through the exercise now.* The account given here is a summary.

The TV programme 'Tawny Owls—a case study' illustrates methods used in Wytham Wood.

Tawny owls are nocturnal and highly territorial; they make characteristic sounds at certain times of year so they can be located and counted. They typically nest in large holes in trees and readily take to suitable boxes, so that the breeding success can be monitored. The chief prey in Wytham Wood are the wood mouse *Apodemus sylvaticus* and the bank vole *Clethrionomys glareolus*; these make up some sixty per cent of the total food, and are the principal food in winter. The populations of these small rodents can be estimated by trapping (see Section 8.5.1). The owls locate their prey at night by sound.

nest boxes

food

TABLE 9 Life table and survivorship data for tawny owls in Wytham Wood.

	in 1952	average for survey
Number of eggs laid	43	62
Number surviving:		
Nestlings in April	16	—
Fledged young in June	15	—
Juveniles that formed pairs in March	9	19
Alive at the end of: two years	—	9
three years	—	7
four years	—	6
five years	—	5

Once a pair of owls has established a territory, they continue to hold it and spend the rest of their lives there; if a female dies, another female will take her place before the following breeding season. Juvenile owls that do not obtain territories move away from the wood and may possibly establish themselves in less favourable places; probably most emigrants die. When the population in the wood was very low, some immigrants from outside were able to establish territories. The life table is given in Table 9.

territory

Southern carried out a key factor analysis, computing the following k-values:

k_1 the number of eggs 'lost' through failure of adults to breed;

k_2 the number of eggs 'lost' through pairs that nested failing to lay the maximum clutch of eggs;

k_3 the number of eggs laid but 'lost' before hatching (usually as a result of desertion by the female);

k_4 the number of nestlings lost before fledging;

k_5 the number of juveniles that failed to establish territories the next spring.

k_1 was the key factor; in 1958 it accounted for the total loss of recruits to the population. k_2 varied in the same way as k_1 but was much smaller; k_3 sometimes varied like k_1 and sometimes differently; k_4 was very low; k_5 appeared to vary in the opposite way to k_1 and k_2.

key factor

Comparison of k_1 and k_2 with the population density of rodents in the preceding winter revealed that the losses due to failure to breed or to lay the maximum clutch showed an inverse relationship to the population density of the rodents below a value of about 20 rodents ha^{-1}; the losses were very high when the numbers of small rodents were very low. Above the limiting density of 20 rodents ha^{-1}, the loss represented a fairly constant low value. So the key factor 'mortality' is related to the general level of abundance of mice and voles in the Wood; it operates through failure of owls to breed when the rodents are few.

The TV programme 'Tawny Owls' shows desertion of a nest by a hungry female in a year when rodent numbers were very low.

ITQ 20 Suggest how scarcity of small rodents could result in owls failing to breed.

Read the answer to ITQ 20.

It is notable that k_4 showed no relation to rodent numbers and was very low every year, implying that the few young that hatch in years with low rodent numbers are then reared successfully.

Figure 10 shows k_5 plotted against the number of young fledged.

tawny owl

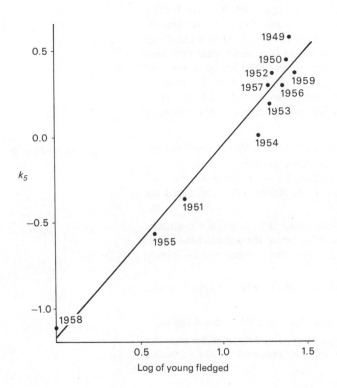

Figure 10 Tawny owls in Wytham Wood: the relation between overwinter disappearance of yearling owls (k_5) and the number of young fledged (expressed as log N).

ITQ 21 What sort of relationship is shown in Figure 10?

Read the answer to ITQ 21.

The net result of adult deaths, emigration of young and (occasionally) immigration is that the population of adult tawny owls at Wytham remains almost constant. Southern's study began in 1947 after a winter of prolonged frost and snow, during which many adult tawny owls probably died from starvation. He therefore started with a small number of pairs (seventeen pairs on the estate); the numbers crept up to reach a steady state of about thirty-one pairs. As the number of the territories increased, the average number of owlets fledged per pair of owls fell from about 1.2 (1947) to about 0.5 (1957), comparing years when there were plenty of rodents in the wood. This is a density dependent effect, with breeding success declining as the number of pairs increases (and, consequently, the size of their territories decreases).

The main regulating mortality factor is k_5, the disappearance of owlets that are unable to establish territories. Given that adult mortality is usually low, there are few vacant territories in winter and the majority of owlets must move out of the area. Only in years when breeding was a disastrous failure were owls able to move into Wytham from outside.

regulating mortality

8.4.2 Populations of the great tit *Parus major*

This bird has been studied in Holland, but the investigation described here was carried out in England at Wytham, near Oxford. It was initiated by Dr. David Lack, F.R.S., in 1947 and has been continued by C. M. Perrins with contributions from various colleagues and research students. Great tits are diurnal (day-feeding) birds and nest in holes; they readily take advantage of nesting boxes of suitable design and this makes it easy to observe and count breeding pairs, and to observe the number of eggs, chicks and fledged young. The population at Wytham seems not to move far away from the Wood, judging from recaptures of marked birds. The provision of nest boxes has probably resulted in an unusually high population of tits and other hole-nesting birds in Wytham Wood.

Great tits feed mainly on insects but take seeds, especially beechmast, in winter; they readily come to bird tables in gardens, especially in winter. The eggs are usually laid in late April, but the date on which the first egg was seen has varied between about 10 April and 10 May, depending on whether the 'season' (observed as time of bud burst of spring plants) has been early or late. The young are fed largely on caterpillars, e.g. of winter moth and green tortrix. The usual clutch of eggs is between eight and ten, but fifteen eggs were recorded for five nests in 1948. The eggs hatch after brooding for about fourteen days and the young take another two weeks to fledge. Both parents bring food to the nestlings and they continue to feed the young tits for about two weeks after they have left the nest. At this stage, parents and young are usually high up in the trees and difficult to observe. Young tits can be caught again about the beginning of November, when young and old birds feed close to the ground and in leaf litter. Pairs form in autumn and begin to take up territories late in the following January.

great tit

The basic census data collected from part of Wytham Wood consisted of: numbers of breeding birds; numbers of eggs laid; numbers of eggs hatched; numbers of young fledged; number of breeding birds in the following year. From more than twenty years of this data, J. R. Krebs (1970) calculated the following k values:

census data and key factor analysis

k_1 is the 'mortality' represented by eggs that were not laid, that is, the amount by which the actual number of eggs counted fell below the potential maximum number if all breeding females had laid 12.5 eggs (this number was the highest average number observed in any year);

k_2 is 'hatching failure', the difference between the number of eggs laid and the number of nestlings hatched out;

k_3 is 'nestling mortality', the number of nestlings that died before fledging;

k_4 is the 'mortality outside the breeding season', the difference between the numbers of young that fledged and the number observed in the following breeding season.

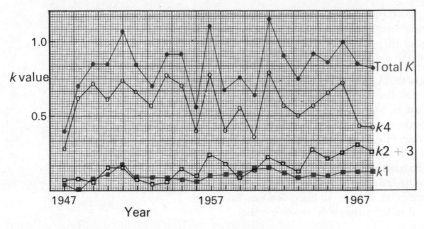

Figure 11 Great tits in Wytham Wood: annual values for $K, k_4, (k_2 + k_3)$ and k_1 between 1947 and 1968. (From J. R. Krebs, 1970.)

ITQ 22 Look at Figure 11. Note that k_2 and k_3 are added together. Which mortality is the 'key factor'? Is there any mortality that clearly looks as though it is a regulating factor, i.e. it is 'density dependent'?

Read the answer to ITQ 22.

For some years it was possible to divide k_4 into two components: $k_{4.1}$ which is the 'autumn mortality', between fledging and the number observed in the winter flocks feeding on the ground; $k_{4.2}$ which is the 'winter mortality', between the numbers in November and the number in the next breeding season. The counts on which these values depend were of a sample of birds, not the whole population.

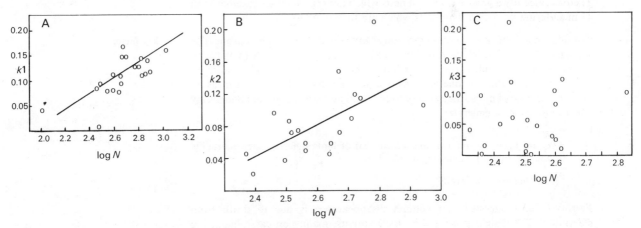

Figure 12 Great tits in Wytham Wood: three k values plotted against the log of the initial populations on which they acted. (A) k_1 against log (maximum clutch size $+ 2$)N. (B) k_2 against log (actual clutch size $+ 2$)N. (C) k_3 against log (number of eggs hatched $+ 2$)N. (From J. R. Krebs, 1970.)

ITQ 23 Look at Figure 12 a, b and c. Which of the k values are density dependent?

Read the answer to ITQ 23.

Expressed simply, (a) the clutch size tends to be larger when the population is low than when it is high, and (b) the mortality between laying and hatching of eggs is lower when the population is low than when the population is high. Thus, as a result, more nestlings are produced per pair of parent birds when the population is low, and relatively fewer nestlings when the population is high. Given that all further mortality is not related to density, this would tend to stabilize the population at an intermediate level.

When the two components of k_4 were plotted separately, neither showed any density dependence.

A computer simulation revealed that the density dependence observed for k_1 and k_2 was sufficient to regulate the population of the study area at a level of about forty-four pairs of adult great tits; the observed fluctuations have been between seven and eighty-six pairs with a mean value of forty-five pairs!

regulating mortality

Perrins (1965) discussed the factors that could affect clutch size and hatching mortality in great tits. He found that the date of breeding (early or late), the age of the female, the nature of the habitat all had some influence on clutch size, but probably the most important factor is the nutritional level of the female, who may produce twelve eggs in a period of twelve days, equivalent to doubling her own body weight in that time. The date of breeding is related to the bud burst of the trees and to the hatching of caterpillars—when the weather is cold in April, all of these tend to be late and, in a warm spring, all tend to be early. Presumably, when the population density is high, females find more difficulty in obtaining sufficient food to lay a clutch of maximum size.

clutch size

The causes of failure of eggs to hatch are infertility, desertion by parents, and predation, mainly by weasels *Mustela nivalis*. Of these three, Krebs demonstrated that only predation by weasels is significantly density dependent. Further analysis of records produced the values in Table 10.

TABLE 10   Predation by weasels on nests of great tits spaced at different densities. (Data from 1958 to 1968.)

Distance of nest from nearest neighbour	Number predated	Number not predated	Percentage predation
Less than 45 m	43	147	23
More than 45 m	34	273	11

There is evidently a greater chance of nests which are spaced closer together than 45 m suffering predation than nests which are further apart.

Consider now the key factor—'mortality' outside the breeding season. The average number of fledglings raised by a pair of adults was six, but, by the time the birds had descended from the treetops, the end of the period of autumn 'mortality', the ratio was one adult to one juvenile. Note that k_4 is a measure of *disappearance* of birds; they are not found dead and it is a point of controversy whether they die or emigrate.

key factor

> **ITQ 24**   From this information, which class of birds suffers greater 'mortality' in the late summer, adults or juveniles?
>
> *Read the answer to ITQ 24.*

Perrins (1965) suggested that autumn disappearance is due to death from starvation. The fledglings are fed by their parents, mainly on caterpillars; the majority of species of caterpillars have finished feeding and have pupated by the time that the young birds become independent. They probably feed on aphids and other small insects high up in the trees; it is possible that they have difficulty in finding sufficient food. Perrins showed that the lighter juveniles and the ones hatched later in the season were those most likely to disappear in the late summer; heavier young which fledged earlier in the summer were more likely to be caught again later.

A possible cause of winter disappearance is starvation; the birds have been observed to spend most of the daylight hours feeding. They are homoiotherms, so they need more food in colder weather; probably many die in cold winters. Perrins estimated that 75 per cent of all the great tits alive in the autumn of 1962 disappeared in the severe winter of 1962–3; for the whole period of observations, disappearance of juveniles varied between 90 per cent (in 1962–3) and 20 per cent or less. The 'winter disappearance' period merges into the setting up of territories for the next breeding season (when the birds are counted again). The number of territories in the study area varied (so their area also varied). Two Dutchmen suggested twenty years ago that winter disappearance was largely the result of emigration by birds that were not able to set up territories, implying that territorial behaviour helps to determine the size of the breeding population.

winter disappearance

Krebs (1971) set out to test this hypothesis, that the presence of territory holders limits numbers in a given area. If, when residents are removed from a stable population, they are replaced by new individuals, this implies that potential settlers had not been able to occupy territories because of the presence of the residents; since territories are prerequisites for breeding, this would mean that territorial occupation limits breeding density. The removal experiments were carried out in a small isolated woodland in which there was a considerable excess of nest boxes. Fourteen to sixteen pairs of great tits bred in this wood each year and almost all the males and most of the females were individually colour-ringed. In 1968, half the pairs there (seven) were shot in mid-March when territories had been established; within six days, five new pairs had settled. The experiment was repeated in the next year: six pairs were removed and four new pairs had settled within three days. All the settlers bred and there were no differences in breeding performance between them and the original residents. Before and shortly after the shooting, the whole area of the small wood was divided into territories among the great tits. There were indications that more birds tried to take up the vacated territories than were able to settle in the end.

The origin of the 'surplus' birds that came in to occupy the vacant territories was investigated in 1969 when nearly all tits within 800 m of the wood were colour-ringed. It turned out that the new settlers came from nearby hedgerows where they had already established territories; these hedgerow birds were in their first year of breeding and it appeared that woodland territories were more 'desirable' than those in hedgerows. The vacated hedgerow territories were not occupied by other birds, implying either that there was no reservoir or surplus population in the area in mid-March or that surplus birds were unable to take up territories then. The success rate of hedgerow nests was lower than woodland nests (20 per cent compared with 90 per cent); the unsuccessful birds were evicted by tree sparrows.

The mechanisms by which tits maintain territories and recognize when territories become vacant are probably through song; very few actual contacts were observed between established residents and intruders. Krebs recorded songs and demonstrated that great tits recognized the song patterns of their neighbours (to which they did not respond as actively as to the song of a stranger). It seems likely that the tits in the hedgerows noted the absence of a particular song pattern when a resident was shot and then rapidly moved into the vacant territory.

Krebs investigated the effect of augmenting the winter food supply by providing sunflower seeds (a good source of protein, acceptable to tits) in a limited area. The birds visiting these were ringed and were observed later breeding in that area but not in a control area where no food was provided. There was no increase in the density of breeding great tits in the experimental area compared with the control area, suggesting that territorial limitation of breeding density is not affected by artificially supplementing the food supply.

Great tits pair before establishing territories, but they normally collect most of the food for the nestlings within their territories. Yet the average size of the territories in any year does not show correlation with the subsequent food supply (caterpillar density) that year. Territory size varies from year to year—how is this size adjusted? From Krebs' analysis of data it seems that territory size is determined by the pressure of intruders; if there is a high population of potential breeders, the territories will be smaller than when there are low numbers of potential breeders. Birds with more neighbours spend more time defending their territories at the time of establishment and it is possible that there is a limiting situation such that the birds spend so much time on territorial activities that they have insufficient time for feeding and breeding. One possible reason for smaller clutches in years of high population density is that males spend extra time defending their territories and so bring less food for the females who consequently lay fewer eggs. Birds which are excluded from territories in optimal areas (woodland) may breed in less suitable habitats such as hedgerows; some birds fail to establish themselves anywhere and emigrate—and die? Once established in a territory, the birds will come back to that area the following year if they survive. But one out of every two adults is likely to die without breeding again.

territory

hedgerow and woodland tits

song

size of territories

8.4.3 Populations of the red grouse *Lagopus lagopus scoticus*

Populations of this bird in Scotland have been studied since 1956. The research emphasized the making of predictions and testing them by experiments and by observing events. The red grouse is a sub-species of the willow grouse, which has a circumpolar range and is typically found on sub-arctic and arctic tundras dominated by a good covering of heaths with patches of scrub willow and birch. In the British Isles, the red grouse lives on heather moors where *Calluna vulgaris* is the principal plant; the birds take shoots of this heather as their principal food.

The TV programme 'Red Grouse— a case study' illustrates the work carried out by Dr. Watson and his colleague Dr. Moss at Kerloch in the Nature Conservancy's Research Station at Blackhall (now part of the Institute of Terrestrial Ecology).

Apart from high ground above 600 m, heather moors are managed habitats, subject to sheep grazing and periodic burning. The red grouse is one of the most important 'sporting' birds in Britain and consequently any marked variation in numbers is of more than academic interest. The research that is the basis of this Section was started because the bags of grouse had fallen; the intention was to find how to reverse this decline and thus provide a scientific basis for improved management.

Grouse show a fairly regular annual cycle of behaviour. Cocks begin to establish territories in autumn and form temporary pairs; birds that do not obtain territories are driven off and move around in flocks, suffering heavy losses from predation. From January or February onwards, in snow-free weather, the cocks spend all day in their territories and pair bonds become stronger, usually lasting until the next autumn. The remaining surplus birds now live near the edges of the heathery ground and nearly all of them die before the breeding season. No pair raises more than one brood per season. After the eggs hatch, the chicks leave the nest within one day and the grouse are no longer confined to territories. The young feed themselves from the time of hatching, at first taking many insects as well as vegetation; by two weeks old, they are already on a diet largely of heather. They remain with their parents for at least three months. Breeding success varies greatly from year to year. The shooting season begins on 12 August by which time the juvenile birds have fully grown wings and tail.

territory, breeding and food

red grouse

Grouse have been marked with rings and also with numbered plastic tabs which can be read with binoculars. With the help of dogs, the research team have been able to count all birds from autumn to spring, to find nests and, later, chicks, and to recover dead birds and dead chicks. Table 11 gives part of the evidence for considering these birds to be mainly those that have failed to establish territories in the autumn.

TABLE 11 Comparison of deaths from predation between 1956 and 1961 of back-tabbed grouse with and without territory (Jenkins and Watson, 1970).

	Number marked	Percentage found killed up to following August
Birds with territories in November	383	2
Birds without territories in November	261	14

Territorial encounters have been observed from Land Rovers, and the territories plotted so that their sizes could be compared. Hypotheses based on observations on the moors have been tested by investigations on captive birds and by experimental alterations of wild grouse populations and certain stretches of moorland. The two main questions were: what causes grouse numbers to fluctuate within areas, and what causes differences in mean population density on different areas?

Tackling the first question by a key factor analysis, Watson (1971) calculated the following k-values:

key factor analysis

k_0 the potential 'loss' due to failure of the adult population to produce the maximum possible number of eggs;

k_1 eggs taken for research purposes—this was often nil and always negligible;

k_2 eggs taken by predators;

k_3 eggs lost as a result of the hen deserting the nest;

k_4 eggs which failed to hatch because of infertility or death of the embryo;

k_5 loss of chicks between hatching and reaching adult size;

k_6 loss due to shooting (August 12 on);

k_7 loss between the end of the shooting period and mid-winter;

k_8 loss between mid-winter and spring;

k_9 loss due to unmated cocks that do not form pairs with hens.

Data were available from several different sites and k-values were calculated for Kerloch, near Aberdeen, and Glen Esk in Angus.

The key factor at Glen Esk was 'winter loss' ($k_7 + k_8 + k_9$) which was closely correlated with K. Plotting this against the population present at the end of the shooting season revealed a delayed density dependent effect, implying a cyclical fluctuation in numbers with the factor influencing the population numbers one or two years later than it acted on the animals. At Kerloch, there was no shooting (no k_6); winter loss was again a key factor, but losses at the egg and chick stages (k_0 to k_5) were also key factors, in some years with greater effect than winter loss; both types of loss showed delayed density dependence.

key factor

Table 12 gives data that support the idea of cyclic changes in grouse populations, but note that the fluctuations are irregular in size and in length of time between successive 'highs' and 'lows'.

TABLE 12   Changes in breeding stocks of grouse related to breeding success. (Jenkins and Watson, 1970.)

Breeding success	Ratio of young produced to adults	Number of areas where subsequent breeding stocks:		
		increased	remained within 10% of original	decreased
Good	>1:1	33	12	8
Poor	<1 or 1:1	3	4	27

Watson, Moss and their collaborators have studied the mechanisms underlying some of these changes in numbers, especially behaviour and nutrition. The cocks hold territories. Differences in territory size within an area can be related to differences in the level of aggression of the individual cocks; a more aggressive cock holds a larger territory. The total number of paired birds in any area in spring depends on the size of the territories that the cocks chose in the previous autumn.

Aggressive behaviour in grouse is shown in the TV programme 'Red grouse—a case study'.

Bringing a sample of eggs off the moor and then incubating them and rearing the chicks in standard conditions in captivity revealed that the proportion of young reared varied from year to year in a similar way to that on the moor. This implies that losses due to k_0, k_4 and part of k_5 are largely determined by factors which affect the hen before she lays her eggs. The chief factor identified is the amount and nutrient content of the heather in the previous spring and winter. Grouse living on a moor that was treated with fertilizer, which increased the growth and nutrient content of the heather, reared larger broods than those on untreated similar moor.

maternal nutrition

In years of population decline, when chick survival is poor, the young cocks that get territories are more aggressive and take larger territories than cocks of older year-classes. In years of population increase, when chicks survive well, the young cocks that get territories are less aggressive and take smaller territories than cocks from older year-classes. Thus the successful cocks' territory size and level of aggression are related to the quality of the eggs from which they hatched.

chick survival

This evidence of an indirect link from maternal nutrition before egg laying to the spacing-out behaviour of the territorial offspring is not the only factor involved in determining territory size. An experiment using fertilizers showed that direct adjustment of territory size to food supply in late summer or autumn can also occur. Fertilizer was spread when the hens were already laying or incubating eggs.

QUESTION   Would you expect the number in full-grown broods reared on the fertilized area to be (a) greater; (b) smaller; or (c) the same as on control (similar) untreated areas?

ANSWER   (c) The same—because brood size is probably determined by the nutrition of the hen before she begins to lay eggs.

The average number in full-grown broods was the same as in the control areas, but in autumn the cocks reared on the fertilized area took much smaller territories than those reared on the control area. This is evidence that better nutrition of the chicks can result in cocks taking up smaller territories (showing lower aggression) even though egg quality, depending on the mother hens' nutrition, had been similar to that of cocks which took up larger territories. The research team's model of grouse population dynamics now involves a direct adjustment relating territorial behaviour to nutrition of chicks as well as an indirect relationship via maternal nutrition and egg quality.

> **ITQ 25** From the information given so far, suggest an explanation for fluctuations in grouse numbers on a Scottish moor.
>
> *Read the answer to ITQ 25.*

For most of its life, the main diet of a grouse is heather tips (shoots and flowers) and usually there are vast amounts of this food on heather moors. However, heather is a poor food by agricultural standards, and the birds, especially hens in spring, are very selective. These observations suggest that only a small proportion of the heather may be a sufficiently good food source to allow the female grouse to produce young which survive well.

Grouse numbers fluctuate from year to year on all moors, but some moors support far higher mean densities than others. Heathery moors support more grouse than wet grassy moors. Those heathery moors over base-rich rocks support higher mean densities of grouse than those over poor rocks. The link here is that heather growing over base-rich rocks has a higher content of 'nutrients' than that growing over poor rocks. On average, the grouse rear larger broods on the nutrient-rich moors and territories are smaller. Two other factors that lead to differences in numbers between areas are visibility and burning. Grouse territories are larger on open ground than on hillocky ground. After burning, the young heather shoots have a higher nutrient content than older heather; the territory size taken by cock grouse decreases after burning, provided that there are still patches of old heather standing, giving shelter and cover.

Calluna vulgaris

8.5 Populations of mammals

> **Study comment** As you read about these mammals, you should think about the problems of observing them and obtaining reliable samples. Voles (Section 8.5.1) are small and cryptic, and observations on different populations give conflicting results. You should compare the lions and wildebeest (Section 8.5.2) with the wolves and moose (Section 8.5.3) as large predators and their prey operating in the very different climatic conditions of East Africa and North America. Although both carnivores form social groups, the organization of these are very different.

Mammals range in size from the tiny pygmy shrew (2.5 g) to the enormous blue whale (1.3×10^8 g), and show a great range in habits and habitats. Interest in fluctuations in mammal populations probably started with Elton's observations on some mammals of the Arctic, especially the snowshoe hare, Arctic fox, lynx and lemmings (Elton, 1924). By using the records of furs bought by the Hudson's Bay Company, he was able to trace fluctuations back to the 1820s. From this there came studies on voles in temperate areas and on various types of mice and rats; these rodents present certain difficulties to investigators because they are usually nocturnal and spend much of their lives in burrows, but they can be marked in various ways and so recognized as individuals.

When asked about conservation, people often think at once of large mammals—elephants, lions, tigers and African antelopes—and there have been a number of studies of populations of ungulates (hoofed mammals), carnivores and other large mammals in connection with the setting up of National Parks and Game Reserves. The large aquatic mammals (whales, dolphins and seals) have been studied because they are of economic value as sources of food and other commodities (sperm oil, oil for margarine, fur, 'bone' for billiard balls, etc.). Many species have been subjected to over-fishing and some of these will be discussed in Block D. There is discussion in that Block also of some of the population problems of *Homo sapiens*.

As case studies in this Unit, there are three examples of mammal populations:

(a) voles, mainly in the USA;

(b) the lions and wildebeest of East Africa;

(c) the moose and wolves of Isle Royale, Michigan, USA.

8.5.1 Vole populations

Voles are small rodents that feed on plants and live in burrows; they may be active by day but are more usually nocturnal. They may breed several times in the year, so the populations consist of overlapping generations. They can be sampled using Longworth traps set in a grid, but problems arise because some animals become 'trap-happy' and others are 'trap-shy'. Comparison of estimates from traps set at different times certainly indicates whether population densities are remaining stable or fluctuating, but it is difficult to give absolute numbers within acceptable statistical limits. Many members of the population do not enter traps, e.g. most breeding females and their young.

Recently voles have been studied by C. J. Krebs and his collaborators in Southern Indiana, USA; the species there are *Microtus pennsylvanicus* and *Microtus ochrogaster*. The *Microtus pennsylvanicus* populations tended to reach peak densities at two-year intervals, and the other species at three- or four-year intervals, but they were all disturbed to some extent by agricultural operations. Figures 13 and 14 show how the numbers of voles changed in the same period on two different grids each of area 0.8 ha. Figure 13 is based on a population allowed to vary naturally over a number of seasons. Figure 14 follows changes when nine females and seven males from a similar population to that shown in Figure 13 were released into a fenced area of 0.8 ha. The numbers entering the traps are assumed to represent a constant proportion of the population in the area; they are not the whole population.

The TV programme 'Tawny Owls— a case study' shows how small rodents are sampled with Longworth traps.

common vole

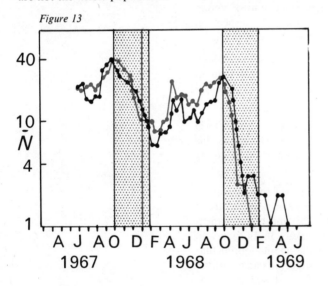

Figure 13

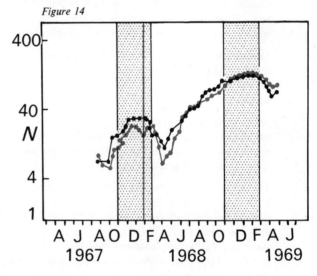

Figure 14

ITQ 26 What are the main population trends shown by Figures 13 and 14?

Read the answer to ITQ 26.

The differences in size of peak populations between the two areas seems to indicate that dispersal (movement away from the area) of voles is a mechanism that might prevent a natural population reaching very high peaks in Indiana. The dense, fenced population overgrazed the habitat; overgrazing was not observed in unfenced populations.

Krebs (1971) summed up earlier work as having shown that population changes can be predicted from values for the reproductive rate (measured as the percentage of trapped females that are lactating), juvenile recruitment (mean numbers of young voles recruited into the trapped population), and female survival (measured from the records of marked females trapped at intervals); there is no need for data about male survival. When a population is increasing, survival of juveniles and females is high and breeding continues through the winter period. In peak populations, juvenile survival and reproductive rates are

Figure 13 Voles *Microtus ochrogaster* in Indiana: changes in population density (expressed as N, derived from the number trapped) on a logarithmic scale in a standard grid in an unfenced area. Red—males; black—females. (From C. J. Krebs, 1971.)

Figure 14 Voles *Microtus ochrogaster* in Indiana: changes in population density (expressed as N) on a logarithmic scale in a fenced area; the population started with nine males and seven females. Red—males; black—females. (From C. J. Krebs, 1971.)

population cycles

31

lower and when population is declining, adult females show lower survival as well as lower reproductive rates, and juvenile survival is also low. These changes in reproductive rate and survival are not observed in fenced populations, in which the numbers rise until the food supply becomes limiting.

Myers and Krebs (1971) monitored the dispersal of voles by trapping and removing all the individuals that moved into two plots in the study area in Indiana; control populations were studied by routine trapping and release. Some tagged animals moved from the control to the 'vacant' areas and this allowed estimation of actual dispersal from the control areas. For *Microtus pennsylvanicus*, when the population was rising, more than thirty per cent of the losses of adults and juveniles could be explained by dispersal from one control area monitored by a vacant area 200 m away but, when the population was declining, the proportion of the fall in numbers that could be explained by dispersal was very low. There was no significant difference in weight range between male voles that dispersed and animals that remained in the control areas, but the females that dispersed were younger and lighter than those that remained.

dispersal

The male voles removed from the vacant areas were given several tests of behaviour in the laboratory; they tended to be more aggressive than resident males during the peak population phase, but the figures were not very convincing. Females were not tested, but differences in aggressiveness and in success in rearing young have been described by other workers.

Myers and Krebs summarized the changes they observed in numbers and condition of the voles *Microtus pennsylvanicus* at different stages of the population cycle:

During the phase of increase of population, the amount of dispersal is high and dispersal accounts for a high proportion of the 'loss' in populations.

cyclical changes in behaviour

During the early stages of the peak population, a high proportion of young females are dispersing, but these are fewer in numbers than the males dispersing, which are mainly older animals with a higher level of aggressive behaviour.

During the late stage of the peak population, the dispersing males have a high level of aggression and the dispersing females are lighter in weight than in earlier phases; all males (resident and dispersing) show a high level of exploratory behaviour.

During the decline phase from the peak population, there was a very low level of dispersal and the mortality could not be explained as being due to this movement; the animals found dispersing included a high proportion of young females in breeding condition.

There were differences in details of the events between the two species studied by Krebs and his collaborators; there are differences between populations in different places and among all the other species of *Microtus* and between them and the bank vole *Clethrionomys*, studied by other workers in Europe and America. Various explanations have been suggested for the population fluctuations of these small rodents including: failure of food supply; changes in the food quality affecting the voles physiologically; depletion of soil nutrients, especially sodium (not supported by evidence collected by Krebs' group); year to year variations in rainfall (and snowfall in places); changes in the physiology of the voles related to population density; changes in behaviour of the voles related to population density (probably correlated with physiological changes); changes in the genetic constitution of the vole populations related to differences in aggressive behaviour and in survival between different genotypes at different phases of the cycle.

> **ITQ 27** Using the evidence given here, which of the possible explanations listed above appear to be promising areas for further research?
>
> *Read the answer to ITQ 27.*

You may have noticed that predation is not mentioned as a possible regulating mechanism. This is justifiable for some vole populations, but it is possible that predators regulate other populations; they may delay increase in numbers after periods of low population.

Myers and Krebs' work seems to demonstrate clearly that mortality during the decline phase is really due to high death rate and is not the result of emigration (which is highest during the increase phase of the cycle). The behaviour and fate of the young females, on whom the ultimate survival of the population largely depends, seems to be a very important aspect of vole populations which has so far not been investigated; when these young females and the juveniles show a high rate of survival the population increases in spite of considerable dispersal. At the peak population stage, juvenile survival, female survival and female reproductive rate all fall and it is tempting to relate this to increased 'social stress' in the dense population, but the evidence for this is still circumstantial and the physiological basis is still unclear.

young females

8.5.2 Lions *Panthera leo* and wildebeest *Connochaetes taurinus* in East Africa

The lion *Panthera leo* is a sociable cat; the population is organized into 'prides' which consist typically of one or more males, several adult females and their cubs. The total number may be anything between four and thirty individuals. The members of the pride may scatter singly or in groups over an area which is their territory; the membership may remain very constant for several years except for births and deaths and emigration of some juveniles. Lions are most abundant in open woodlands; it seems that prides are probably smaller in habitats where trees are more dense. Schaller (1972) has reported on lions in the Serengeti area (Tanzania) and reviewed other information and studies.

lion and lioness

Lionesses are the central group of the pride; they are all related to each other. Adult males are members for periods between a few months or six years. The lionesses are aggressive towards other lionesses that are not members of the pride, but may court nomadic males or males of neighbouring prides. All the adult lionesses may breed and several litters may be born at the same time; their care is then a communal affair and the cubs may suck milk from any lactating female, not just from their own mother. Female cubs remain with the pride if they survive to become adult, but males typically leave before they become sexually mature at about three-and-a-half years old. The adult males of the pride are often brothers, or grew up together, or became companions before they joined the lionesses.

The lions are inefficient hunters and more than 85 per cent of the hunting of the pride is done by the lionesses. Lionesses usually hunt together and typically they hunt prey of 100 kg or more, e.g. wildebeest, zebra, topi, giraffe or buffalo. When individuals hunt alone, they stalk smaller animals, and this happens when smaller animals are the most abundant prey available. A group of five or six are able to gorge themselves on the prey and consume most of the carcass within a couple of hours, but sometimes one member guards it until others arrive and sometimes food is taken back to members guarding cubs. Males run up when the kill has been made and then gorge themselves, sometimes to the extent that cubs die of starvation. However, lions may take remains of kills and prevent lionesses from eating while allowing cubs to feed; this behaviour may enhance the survival of cubs. The lionesses suckle, guard and transport the cubs for the first few weeks of their lives, and continue to protect them and provide food until they are almost independent, but they may keep starving older cubs away from food until they have fed themselves.

food

The territory over which pride lionesses wander varies between twenty and four hundred square kilometres; nomads wander over much greater areas, four thousand square kilometres and more. The territories are marked by scent sprays.

territory

ITQ 28 Suggest some possible advantages of the territorial habit to lions.

Read the answer to ITQ 28.

The communal habit appears to result in better reproductive success, especially if pride males are present; nomadic females produce cubs but seem to rear very few of them. Schaller (1972) reported that a normal pride with vigorous lions present reared 46 per cent of the cubs born up to one year old, but another pride, which had lost the pride lions, reared only 19 per cent of the cubs.

Lions have a high reproductive potential; the gestation period is about four months and lionesses that lose cubs can come into oestrus within a few days

33

and produce another litter only four months later. If at least one cub is reared, reproductive potential the interval before the next litter is eighteen months or more. Assuming an average litter size of 2.3 and a birth interval of two years, a lioness should be able to rear an average of 1.2 cubs per year, but few actually achieve this in the Serengeti where only 23 per cent of the expected total of cubs was raised during Schaller's period of study. Two contributing factors were that about 15 per cent of the adult lionesses appeared to be barren, and the average interval between the death of one litter and birth of the next was twice as long as expected. At least causes of death two out of every three cubs born died before becoming independent; the causes included abandonment (possibly half the cubs), predation (by hyenas and others), violence from adult lions, and starvation. In an area where the food supply was not limited, about half the cubs died before becoming independent. The annual death rate of pride lions was about 5.5 per cent; the causes of death were old age (fifteen years counts as old, but one female is known to have lived to twenty-two), violence (usually due to human activity or from other lions but sometimes from other larger animals) or illness. Healthy adult lions probably never die from starvation in the Serengeti, but injured or sick animals may starve to death and many cubs starve at the stage when they become too large to obtain much milk but are too small to hunt or to compete successfully at kills. The annual recruitment of juveniles observed by Schaller was about 11 per cent, which is twice the annual death rate of adult lions.

> **ITQ 29** Suggest how it is that the average numbers in prides remain constant in spite of this excess of annual recruitment over annual death rate of adults.

Read the answer to ITQ 29.

One of the points of discussion about lions and other predatory mammals has prey biomass been their relationship with their prey. Consider the evidence relating the size of lion territories to the density of their prey, the large ungulates. Compare the figures in Table 13.

TABLE 13 Prey abundance and lion density in five African reserves.

Reserve	Number of lions	Area per lion (km²)	Biomass of prey (kg km⁻²)	Number of prey present per lion
Ngorongoro	70	3.7	10 363	338
Manyara	35	2.6	7 785	70
Serengeti	2 200	11.6	4 222	437
Nairobi	25	4.6	3 052	152
Kruger	1 120	17.0	1 034	249

> **ITQ 30** Assume that an 'average lion' (including cubs) eats a total weight of 2 500 kg of prey in a year; calculate the proportion of prey biomass that is consumed by lions in each of the five reserves in Table 13.

Read the answer to ITQ 30.

There seems to be a contrast between Manyara, Nairobi and Kruger (where the prey biomass available seems to be six to eight times the average food consumption) and Ngorongoro and Serengeti (where the biomass available seems to be fifteen and twenty times the average food consumption). The figures seem to imply that Ngorongoro and Serengeti could support twice or three times the number of lions that are actually present.

Schaller notes that average densities of lions over large areas never seem to

TABLE 14 Predator biomass in five African reserves as kg km⁻².

Reserve	Total predators	Lions	Spotted hyenas	kg of prey kg⁻¹ predators
Ngorongoro	96	28	68	108
Manyara	45	38	4	174
Serengeti	15	8	5	282
Nairobi	32	21	4	94
Kruger	10	6	3	100

exceed one lion per 2.6km², so that the Manyara reserve carries a maximum density possibly imposed by a spacing out requirement related to social structure.

Another factor not so far mentioned is that the prey biomass is also the food supply for other species of predators. Look at Table 14.

> **ITQ 31** Comment on a striking difference between conditions in Ngorongoro and the other reserves in Table 14.
>
> *Read the answer to ITQ 31.*

The hyenas *Crocuta crocuta* in Ngorongoro form permanent 'clans' with their own territories and they hunt together and are able to prey on large animals such as wildebeest; it seems probable that they are close to the limit of their food supply there. On the Serengeti, the clans dissolve when food is scarce and there is a high mortality of hyena cubs as a result of starvation; the cubs rely on their mother's milk for eight months and so are very vulnerable to reduction of adult food supply.

spotted hyena

game migrations

Serengeti thus stands out as apparently having a lower density of lions than would be expected from the biomass of prey. The explanation lies in the habits of the ungulates contrasted with the habits of the lions.

The Serengeti is famous for the spectacular migrations of the 'game'. More than half a million wildebeest and zebra feed on the grass and their movements are related to the seasonal rains and the flush of vegetation that provides a nutritious food supply. As the plains dry up and the vegetation dies back, the herds move to other areas. Some of the predators follow them, but the pride lions are sedentary in habit and the lionesses remain for their whole lives in the pride's territory. On the plains, the biomass of prey drops below 100 kg km⁻² in the dry season and very few prides occupy territories there; in woodlands, the biomass remains at about 1 000 kg km⁻², even at its minimum, and the whole area is shared among prides. Schaller concludes that the size of the territories of Serengeti lion prides is related to the food supply at the time of year when it is minimum, probably to the level of prey at times of severe food shortage, which recur at intervals of several years. To give you some idea of the totals involved, migratory prey have an average biomass of 70×10^6 kg in the Serengeti area compared with 0.4×10^6 kg of resident prey on the plains and 40×10^6 kg of resident prey in woodlands.

Consider the populations of wildebeest *Connochaetes taurinus* as an example of the ungulates which are the prey of lions and other predators in the African reserves. Kruuk (1970) calculated that 45 per cent of the wildebeest calves died in 1966; the loss might have been greater except that there is a sharp peak in breeding so that large numbers of the vulnerable young calves are produced at the same time and the habit of travelling in vast herds means that there is greater protection for calves. The habit of calving on the plains means that lions and leopards, which both are more characteristic of woodlands, make less impact on wildebeest calves than if these were available in woodland. The 45 per cent loss is partly due to predators such as hyenas and wild dogs but also results from newborn young losing contact with their mothers in large, dense herds; mothers will suckle only their own young, so lost calves are doomed to starve unless eaten first. Estimates of the numbers of yearlings in the population were 14 per cent between 1963 and 1966 and 10 per cent in 1969, and the annual death rate of adults was estimated to vary between 4 per cent in 1962–3 and 16 per cent in 1965–6.

wildebeest

The two main predators on wildebeest are lions and hyenas. The latter prey heavily on males and on sick and starving individuals; Kruuk estimated that the hyenas would take annually 1.7 to 2.8 per cent of a population of about 410 000 animals.

Serengeti

Lions prey on wildebeest for about four months of the year during which time wildebeest make up at least half their food; thus wildebeest probably contribute at least 20 per cent to the total food intake of a lion, giving a total for the Serengeti of about 11 000 animals (including calves), again between 2 and 3 per cent of the population. Many thousands of individuals must die from causes other than predation, probably malnutrition and disease. Wildebeest populations increased dramatically in the 1960s, showing that predators were not cropping them to a low level.

Conditions in Ngorongoro are quite different. This reserve is in the crater of an extinct volcano and there is no possibility of the animals migrating out of it; they are all permanently resident there. The hyenas, which make up the bulk of carnivores, prey on wildebeest heavily, and Kruuk has calculated that mortality due to predation is about equal to recruitment of juveniles. About 75 per cent of wildebeest cows produce a calf in their second year whereas, in Serengeti, only 35 per cent of cows in their second year produce calves, and this figure may have fallen to 4 per cent recently when the wildebeest population increased.

Ngorongoro

ITQ 32 What sort of mechanism does this information about wildebeest cows suggest?

Read the answer to ITQ 32.

Nairobi Game Park illustrates another of the possible relationships between predator and prey. This reserve is small, but the animals are free to move outside it. Wildebeest concentrate in the Park during the dry season, and several thousand were counted each year until 1961, when there was a severe drought which resulted in high mortality. The counts for the following years were:

Nairobi Game Park

1961	1780
1962	956
1963	691
1966	253

Lions before 1961 took wildebeest as prey more than twice as often as would have been predicted if they were taking large ungulates in proportion to the populations present in the area. In 1966, the lions were taking wildebeest four times as often as would have been predicted. Foster and Kearney (1967) calculated that the annual decrease in numbers of wildebeest could be explained entirely by lion predation. From 1967, lions turned more to eland and the decline in wildebeest numbers stopped.

ITQ 33 What sort of response were the lions making to the decline in wildebeest?

Read the answer to ITQ 33.

8.5.3   The moose *Alces alces* and wolves *Canis lupus* of Isle Royale

Some aspects of the moose and wolf populations of Isle Royale are discussed in Unit 3, Section 3.4.4; you should find it helpful to read that Section again now. The island is situated in Lake Superior and is part of the state of Michigan, USA.

The situation on Isle Royale is of great interest because it has developed 'naturally' (i.e. unaided by man). Moose reached the island about 1908; wolves arrived there about 1948. This is probably the only site where this large herbivore and its natural predator still co-exist unhunted by man, and it appears that the populations are in balance with each other and with the vegetation which is the food of the moose.

The moose, *Alces alces*, called 'elk' in Europe, is the largest living member of the deer family Cervidae; the adults are about the size of horses and weigh about 450 kg, and the calves are 1 m tall and weigh about 14 kg at birth. Moose are herbivorous and typically browse on leaves and branches of trees, but in summer they also feed on submerged aquatic plants; they are good swimmers and can dive to feed on plants near the bottom in about 2 m of water. The moose probably reached Isle Royale by swimming from the Canadian mainland which is 15 miles away at the nearest point; some may have crossed on ice in 1912 to 1913. Moose were very abundant in that part of Canada in the early 1900s but were not recorded from Isle Royale by visiting naturalists in 1905.

food of moose

The figures in Table 15 are from different authors and there is doubt about how accurate they really are. In summer, the moose population is dispersed among the woodland, bogs and lakes of the island and it is difficult to observe and count them, even from the air. Mech, who worked on the wolves and moose between 1959 and 1961, estimated that there was a stable population of 600 moose during those years, but Jordan *et al.* (1971) suggest that the stable population between 1959 and 1969 was actually 1000 moose and believe that Mech's

moose population changes

TABLE 15 The estimated number of moose on Isle Royale in different years.

1915	200
1917 to 1918	300
1919 to 1920	300
1921 to 1922	1 000
1925–1926	2 000
1928	1 000 to 5 000
1930	1 000 to 3 000
1936	400 to 500
1943	171
1945	510
1947	600
1950	500
1957	300
1959 to 1961	600 (Mech) or 1 000 (Jordan *et al.*)

moose

method led to systematic underestimations; probably some of the values for earlier years are also underestimates.

Naturalists visiting the island in 1934 found 40 dead moose on the area which they examined (one-tenth of the total); the carcasses were emaciated. They reported that there was no browse left suitable for moose in winter. The ground hemlock *Taxus canadensis* (a yew) had been exterminated and water lilies and pondweeds (eaten in summer) had almost disappeared from the small lakes. In 1945, after the population had 'crashed' and then recovered, the winter browse was described as just adequate; in 1948, it was said to have deteriorated again but, by 1953, the food was said once more to be adequate. It was concluded that the plants on the island could support about 500 to 600 moose, but higher numbers represented overcrowding and overgrazing. Murie (1934) suggested that the moose population should be controlled either by permitting hunting (by sportsmen) or by direct culling or by capture of individuals and transport back to the mainland; in fact, 71 moose were taken to mainland Michigan between 1934 and 1937. Two pairs of zoo-bred wolves were taken to the island, but they proved such a nuisance (because they were 'tame') that they were removed.

TABLE 16 Estimates of the numbers of wolves on Isle Royale in different years.

Year	Minimum number	Best estimate of number
1952	2	
1953	4	
1956	14	
1957	15	
1959	19	20
1960	19	22
1961	20	22
1962	22	23
1963	20	20
1964	25	25 + 1 found dead
1965	25	28
1966	21	23 + 2 found dead

wolf

wolf packs and breeding

The wolves probably arrived by crossing the ice in 1948; there may have been only one pair who eventually set up a pack with their offspring. In 1959, there was one large pack of about 15 individuals and two small packs of two and three individuals (see Table 16). The timber wolf *Canis lupus lycaon* is one of the largest predatory animals of the north temperate zone; adults are about 1.2 m long and weigh 25 to 35 kg. They start to breed when two years old (females) or three years old (males), and the average litter is about six pups, which are born in an underground den and are completely helpless at birth. Wolves are social animals, forming packs which usually include almost equal numbers of males and females. These packs are related individuals, except that the 'alpha' pair, the dominant male and female, may not be related; the rest of the pack probably includes their offspring of various ages. There may be other adult males and females, juveniles of both sexes and pups and also one or two 'senile' individuals (10 to 14 years old). The pack hunt together and are thus able to bring down moose which are much larger animals than the individual wolves. Moose can defend themselves with their hooves. Mech watched 77 incidents when wolves 'tested' adult moose and only 6 of these were killed, giving a predation efficiency of about 8 per cent. The wolves eat either calves in their first year or moose which are more than six years old; moose between one and six years old are seldom killed by wolves, suggesting that they are perhaps more alert and agile than their elders.

Mech observed the wolves intensively for short periods and found that the large pack killed one moose about every three days. He calculated that all the wolves on the island probably took 142 calves and 82 adults each year. The wolves have the habit of gorging when food is available and can take up to 9 kg of meat in a meal. As is typical of carnivores, they can go for several days without food. Adults regurgitate semi-digested food to cubs which stay near the den until they are large enough to join the hunt. Food is also taken back to a female with young cubs who stays with them and suckles them. At the kill, the dominant wolves feed first and wolves low in the hierarchy may go hungry.

food of wolves

Jordan *et al.*, starting with a higher estimate of the moose population, calculated that the wolves took 124 adults. Data about the production of calves is given in Table 17.

moose mortality

TABLE 17 Data about moose on Isle Royale (averages based on 10 years' observations).

Year class	Percentage of total number	Number of females	Number of calves produced
1	12.5	63	0
2	10.8	54	8.8
3 and older	76.7	383	432.8
Totals	100	500	441.6

ITQ 34 Calculate the mortality rate (a) of calves during their first year and (b) of animals during their second year of life.

Read the answer to ITQ 34.

Of the 316 calves which die in their first year, many die from drowning, and probably their bodies are not consumed by wolves; wolves probably account for 200 or more. Mech observed that the rate of twinning among moose in 1959 was 38 per cent and the rate was the same between 1961 and 1963; this contrasts with a twinning rate of 6 per cent observed by Murie in 1929–30.

ITQ 35 What does this change in twinning rate suggest about the moose populations of 1929 and 1959?

Read the answer to ITQ 35.

Look at Table 16 (numbers of wolves); it appears that the total number of wolves on Isle Royale has remained remarkably constant, with an average of about 24, since 1960.

wolf reproduction

QUESTION Given that a female wolf lives for an average of 10 years and starts to breed when two years old (in her third year) and produces a litter of six pups each year, how many pups should the female produce during her life?

ANSWER She should breed 8 times and produce 48 pups.

QUESTION If a wolf lives for 10 years on average, what should the percentage turnover of the population of adult and yearling wolves be, if the population numbers remain constant?

ANSWER About 10 per cent each year.

ITQ 36 What is the probable rate of recruitment of yearling wolves to the population after 1960, assuming a constant population of 24 wolves?

Read the answer to ITQ 36.

The annual recruitment to the Isle Royale wolf population is about half a litter each year on average. At any time, there must be five or six adult females capable of breeding, so the recruitment is of the order of 10 per cent of the potential production of pups. The low rate of recruitment observed at Isle Royale is characteristic of wolf populations in general; it is the result of a low birth rate and a low survival rate. About 40 per cent of adult females do not appear to breed at all and many bear a smaller number of young; often it is the dominant female only who produces a litter. In February, there is a great deal of courtship behaviour among the pack but few copulations are observed and the fertility rate seems to be low. The survival rate of pups between birth and five to ten months old has been recorded as between 6 and 43 per cent; survival rises to 55 per cent for pups between ten months old and reaching the breeding age of about two years; after that, survival is generally of the order of 80 per cent unless there is heavy mortality from hunters.

wolf rate of recruitment

ITQ 37 Construct a possible life table for the Isle Royale wolves assuming that there are three packs, each with a dominant female who produces one litter.

Read the answer to ITQ 37.

In fact, Mech did not observe signs of any pups being produced in 1959 or 1960; he thought that there might have been three yearlings in the large pack when he started work in 1959.

Now you could attempt SAQ 5.

9.0 Introduction

Study comment In the last twenty years, there have been arguments about how animal population numbers change, based on three different theories. That of Andrewartha and Birch (1954) you have already met in *IPE*; the others are those of Lack (1954) and Wynne-Edwards (1962) and both were based on studies on birds. You should try to understand the main points of these theories so that you can apply them to the various groups of organisms in the Sections that follow.

In studies of the numbers of individuals in populations, animal ecologists have been much more active than plant ecologists. The latter have studied the distribution of plant species and have been able to name communities after those species which are either most abundant or most 'characteristic'; counting different stages of the life cycle of plants and setting up life tables is becoming more fashionable but, at present, there are few good studies available.

Animal ecologists have been aware for many years of two problems related to numbers of individuals, though the two have often been confused.

1 Some species are common and others are rare; in a stable environment, the common species remain common and the rare species remain rare.

common and rare species

2 For many species, the numbers in a population fluctuate; the numbers in one generation (or age class or cohort) are markedly different from those in preceding generations (age classes, cohorts).

population fluctuations

Some of the fluctuations in numbers are large and make a dramatic impact on the environment, so much more attention has been focused on this aspect of numbers than on the problem of why the level about which fluctuations occur is very different for common and for rare species.

Andrewartha and Birch (1954) in *The Distribution and Abundance of Animals* argued that the populations of animals are limited in three ways:

Andrewartha and Birch

(a)   there are shortages of resources such as food, shelter, places to breed;

(b)   these resources may be inaccessible to animals because they may lack the capacity to disperse and search for them;

(c)   fluctuations in total numbers may be caused by weather or by predators or by any other environmental factor which influences the rate of change in numbers. This rate of change includes recruitment to and losses from the population, and is not always equivalent to natality or birth rate.

Lack (1954) in *The Natural Regulation of Animal Numbers* put forward the general thesis that populations fluctuate within limits about a medium level and this fluctuation must be regulated by density dependent processes. He developed his ideas further in *Population Studies of Birds* (1966) where he summed up his views as:

Lack

(a)   the reproductive rates of birds have evolved through natural selection and are as high as the birds' capacity and the environment allow;

(b)   mortality rates balance reproductive rates and the numbers in bird populations are regulated by density dependent mortality;

(c)   starvation outside the breeding season is much the most important density dependent mortality factor in populations of wild birds;

(d)   breeding pairs of birds are broadly dispersed in relation to food supplies through various types of behaviour 'which are little understood but which are to be explained through natural selection'.

Although Lack confined the examples and most of the discussion in his second book to birds, he believed that his views could be extended to cover other animals, with the proviso that the most important density dependent mortality factor might be different for different species (which is also true for birds).

Wynne-Edwards (1962) in *Animal Dispersion in relation to Social Behaviour* suggested that animal species typically control their own population densities and keep these as near as possible to what he describes as the 'optimum level' for each habitat. He argued then and later:

Wynne-Edwards

(a)   the availability of food must set an ultimate limit to the size of populations;

(b)   natural populations rarely reach this limit (when over-population would lead to mortality from starvation);

(c)   the animals' behaviour results in dispersion of some individuals and the maintenance of numbers such that those remaining in the area have plenty of food and other requisites. The level of a population remains well below the limit set by food supply;

(d)   the particular behaviour involved in homeostatic regulation of numbers is reponse to 'epideictic' (derived from the Greek for sample-presenting) displays which can be identified by studying the patterns of behaviour of each species;

epideictic behaviour

(e)   the animals react to the level of population revealed by the epideictic displays by restricting birth rates; this and the dispersal behaviour ensure that the population does not increase far above the optimum for the environment;

(f)   the mechanism of population regulation via epideictic displays and restriction of birth rates must have evolved as a result of 'group selection', which is a different process from natural selection of individuals. The argument about whether group selection occurs or not falls within the field of evolutionary biology and is outside the scope of this Course.

Wynne-Edwards drew examples from the whole animal kingdom, but a very large number of them were species of birds; Lack (1966) criticized Wynne-Edwards' theory on the basis of information assembled from studies of birds.

Andrewartha and Birch are both entomologists and the majority of their examples are from studies of insects (refer back to *Insect Population Ecology* for a discussion of their views). They differ from both Lack and Wynne-Edwards in believing that there is no medium level about which numbers fluctuate, so there is no need to postulate density dependent factors that affect animal numbers. Andrewartha and Birch were seeing only half the problem—they were discussing only key factors which affect the numbers present at any one time and often are the results of weather; they ignored the fact that some insect species are rare and others common. Moreover, most of their examples came from insects living

unstable populations

in unstable environments. There is much evidence that populations in stable environments fluctuate about definite medium levels and the regulation of their numbers must involve density dependent factors.

Lack and Wynne-Edwards both believe in density dependent regulation of animal numbers, but they differ as to how this is achieved. Lack believes that birth rates have been subjected to natural selection and the number of offspring produced is the number that can be expected to give rise to the maximum number of juveniles under the average of that particular range of environmental conditions. The size of the breeding population is regulated by density dependent factors, often mortality related to food scarcity. Wynne-Edwards believes in contest competition leading to dispersal and also that the birth rate is controlled by behaviour of the animals in a density dependent way, so that the population is maintained at a level at which there is usually no disastrous shortage of requisites.

density dependent regulation

Now you could attempt SAQ 3.

9.1 Population regulation in flowering plants

Study comment The discussion is based on a population 'model'; you should note the parts played in this by the seed in the ground and factors that affect its germination, the seed produced by the plants and the 'constant yield' principle operating during growth. Look back to Section 8.2.1 for information about buttercups.

Plants of economic importance are those used for agriculture, including horticulture, and those used in forestry; most of the agricultural plants are annuals (e.g. cereals), whereas forest trees are perennials, usually taking many years to reach economic value. So there are likely to be differences in strategy between agriculturists and foresters, but both are interested in achieving optimal 'crop' density, i.e. the greatest return per unit area of land. There is much less interest in plant numbers and how they vary since an increase in numbers may be accompanied by decrease in average yield per plant so that the same crop may come from a range of plant densities.

Harper and White (1971) have proposed a model that 'may serve as a beginning of attempts at more complex model building in more sophisticated systems'. The parts of the model are:

1 a seed reservoir (viable seed including some that may be dormant);

2 an environmental 'mesh' or 'sieve' which determines the number of seeds that germinate and become established seedlings;

3 a stage of plant development during which the plants grow and reproduce within the limits set by environmental constraints;

4 the seeds produced by the plants (which may replenish the seed reservoir or may germinate directly).

QUESTION From your knowledge of buttercups, suggest one feature of the life of plants which may not be covered in this scheme.

ANSWER Vegetative reproduction. Perhaps this could be included as part of the stage of plant development; perhaps it ought to be given a separate status.

Vegetative reproduction is common among perennial plants but its significance in population regulation has been studied much less than the role of seed production.

There has been a tendency to overlook the seed reservoir (because it is buried); populations of up to 75 000 viable seeds m^{-2} have been recorded in arable soils and up to 20 000 m^{-2} in meadow communities. It is interesting that this population of seeds may not correlate well with the species of plants growing in the habitat. It appears that a disproportionate number may be annuals and some may come from outside the area or represent species that lived there in earlier years. For plots protected from fresh seed 'rain', there is typically a rapid exponential decrease in the number of viable seeds of any species present, but a fairly constant proportion of these are likely to germinate each year. Dormancy

seed reservoir

may continue for many years in some species or the seed may 'last' for two or three years only, even if not eaten or otherwise destroyed by that time.

ITQ 38 From your study of Section 8.2.1, suggest a feature that might be included in the 'environmental sieve' for buttercup species.

Read the answer to ITQ 38.

Experiments have shown that minute details of the soil surface (the microsites) may be important for germination of some species. The depth to which the seed is buried and the temperature of the soil may affect germination and the presence of plants either of the same species or of others may promote or prevent germination. Some of these effects of the physicochemical and biotic environment will be taken up in Block C.

germination

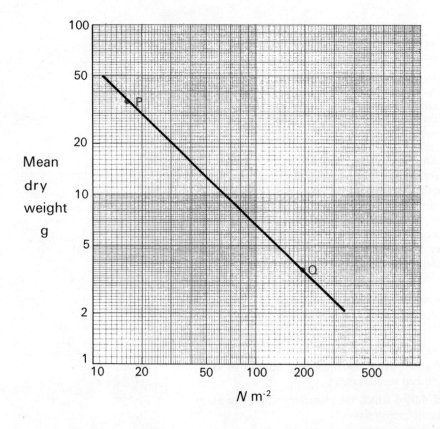

Figure 15 Mean dry weight (g) of soya bean plants harvested after 84 days plotted against the initial density of the seedlings (N m^{-2}).

One of the fundamental quantitative principles of plant dynamics seems to be that, over a wide range of densities, the final yield of plant matter per unit area is constant for many species. This is achieved during the stage of plant development. Look at Figure 15.

constant yield principle

ITQ 39 State in simple words the differences between the plants at point P and point Q (which are the same age and derived from the same seed bank).

Read the answer to ITQ 39.

The adjustment of yield to density which this principle implies may be achieved in two ways: by differential growth so that crowded plants are all smaller than less crowded plants, or by differential mortality so that the same number of plants survive per unit area, provided that the starting density exceeds a threshold value (these plants should be the same size when ready for harvest). In practice what happens is a mixture of these two methods of adjustment. The mortality that leads to reduction of numbers in a crowded stand is called 'self-thinning'. It is well known in forest nurseries and has been demonstrated for many other plants. The 'suppressed' individuals do not always die but may survive for many years exhibiting almost no growth unless there is some change in environment which releases them from suppression; when not growing, they cannot reproduce so that they will not contribute to seed production. Competition among plants (of which this is an example) is treated in more detail in Block C (Unit 13).

self-thinning

The TV programme 'Tropical Forest' shows suppressed trees growing in the shade of mature trees.

The 'constant yield' principle usually affects seed production of the population, as well as plant growth, but sometimes there may be an optimum density for reproduction and more crowded plants may fail to flower and set seed.

ITQ 40 What type of relationship between plant numbers and seed numbers is this?

Read the answer to ITQ 40.

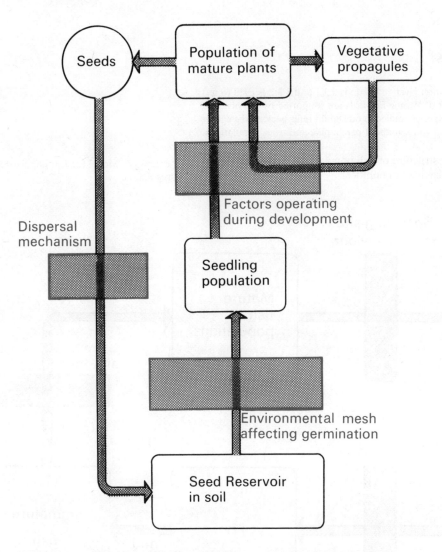

Figure 16 Model for plant population regulation.

The model suggested by Harper and White (1971), shown as a modified version in Figure 16, can be summarized simply:

There is a reservoir of seed in the soil; some proportion of this germinates in any growing season, some of the rest loses viability but some may germinate in later growing seasons. The proportion or numbers germinating are determined by conditions in the soil environment which vary with the species and may include influences from plants of that species growing there. There is likely to be mortality among the seedlings and the growth and reproductive success of the plants may vary greatly, but the principle of constant yield implies that this phase of population regulation is density dependent. The number of seeds produced by the population will thus be less variable than the number of seeds germinating. This seed 'rain' joins the seed reservoir and the cycle starts again. For plants that reproduce vegetatively only, the stage of plant development is the central feature of the model and determines the number of propagules that join the population.

For perennial plants, with overlap of generations and continued seed production over a period of years, analysis of populations is very difficult (compare this situation with population analyses of most vertebrates). Some plants have regular biennial life cycles, reproducing once and then dying; these can be treated as annuals operating over two years, but with two distinct populations present at any one time (the age-class just germinated and the age-class that will flower in this season).

perennial and biennial plants

43

The division of the regulatory system into four stages allows separate analysis of the factors affecting each stage and comparison between species.

ITQ 41 What is the fundamental difference between higher plants and most animals that is emphasized by this model?

Read the answer to ITQ 41.

9.2 Population regulation in fishes

Study comment The information given in Sections 8.3.1 and 8.3.2 is used to test whether the views of Lack or of Wynne-Edwards are supported by the observations made on the three fish species studied. You should refer back to the earlier Sections if you need to remind yourself about perch, pike and trout populations.

Figure 17 shows a model of the structure of fish populations. Stages are shown at which there could be marked changes in numbers.

Figure 17 Model for fish population regulation.

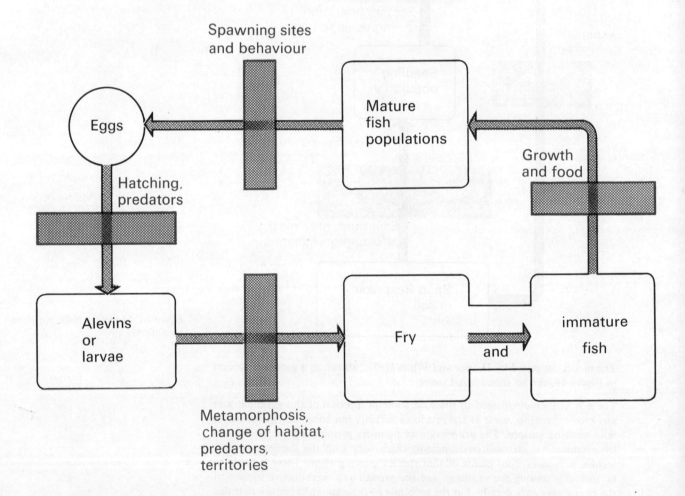

ITQ 42 Recall from the Introduction (Section 9.0) the views of Lack and of Wynne-Edwards about population regulation and state at which places in the model there should be regulation according to their theories.

Read the answer to ITQ 42.

ITQ 43 Is there any evidence from the examples studied in Unit 8 that fish regulate their egg production according to the population density? If your answer is yes, quote the evidence.

Read the answer to ITQ 43.

You were given no information about pike egg production in Section 8.3.1. There is much variation between individual fish. Kipling and Frost (1969) reported that there had been a change in average fecundity between 1964/65

egg production

and 1965/66 from about 27 000 eggs kg⁻¹ to about 30 000 eggs kg⁻¹ (and the difference is very significant at the 0.01 level). This change in level of egg production could have resulted in 5×10^6 more eggs being laid in 1966 than would have been laid at the lower level of fecundity. There are few figures for pike before 1964; counts for three fish in 1950 gave about 20 000 eggs kg⁻¹ and it is very likely (but not proven) that mean fecundity was lower then than in 1964. The implication of these observations is that the mean fecundity of the pike in Windermere may have risen following the reduction in numbers of adult fish; this would support Wynne-Edwards' hypothesis that egg production should vary inversely with density.

perch fry

trout alevin

There is much difference in size of eggs between species of fishes, but within a species the egg size is relatively constant. The implication of laying very small eggs is that the tiny fish which hatch out must very soon feed themselves and require a supply of tiny food organisms (as with perch and pike). If the eggs are large, the fish which hatch out are relatively larger and can start to feed on larger organisms; their period of development within the egg is relatively longer and this allows for a gap in time between the spawning season and the hatching of the eggs. Trout usually spawn between November and February, but the eggs hatch between March and May, having spent the winter season buried in gravel.

> **ITQ 44** Consider the evidence in Unit 8 for mortality outside the breeding season. At which stage does it appear that mortality occurs which regulates the population numbers for trout, perch and pike?
>
> *Read the answer to ITQ 44.*

Trout numbers appear to be regulated by the behaviour of young fry. The trout population is adjusted to the 'carrying capacity' of the river if this is defined in terms of topography of the substratum. Similar adjustment of fry populations to river topography has been shown for some other species of salmonid fishes, e.g. salmon *Salmo salar*.

fry numbers

For pike and perch, the adjustment of numbers is probably related to food supply at a particular stage of the life history and time of year. The adult populations are characterized by 'dominant year-classes'. For both these species, there is evidence that adult populations in some waters may be so dense that the growth of the fish is stunted as a result of restriction of food supply. Populations in these waters consist of many individuals which are relatively small but old, in contrast to the pike population of Windermere, where the fish are small but young.

> **ITQ 45** How will the egg production of stunted populations compare with others?
>
> *Read the answer to ITQ 45.*

After the high mortality of the fry stage, the mortality rate for fishes often seems to remain constant for many seasons. We still know surprisingly little about when and why these older fish die. Some are caught by fishermen or eaten by predators, and others may die of disease or from the effects of a load of parasites. Since fish are capable of growth throughout their lives, older fish are generally larger and are generally less likely to be taken by predators (other than man) than younger fish. Older fish presumably are likely to carry a greater load of parasites; they need a greater supply of food for maintenance. Large numbers of dead adult fish are sometimes observed during the spawning season or as a result of catastrophes such as sudden pollutions. Many fish die and are never observed as corpses, so that there is still mystery about the major causes of mortality of older fish of many species.

mortality of older fishes

Now you could attempt SAQ 4 if you have not already done so.

9.3 Population regulation in birds

Study comment You should study this Section carefully since Lack and Wynne-Edwards have both based their theories largely on data about birds. Refer back to Sections 8.4.1, 8.4.2 and 8.4.3 for the detailed information about tawny owls, great tits and red grouse. Extra information is given about shearwaters and albatrosses and about wood pigeons.

Figure 18 shows a model of the events in the life history of birds and indicates stages at which there could be increases and decreases in population.

Figure 18 Model for bird population regulation.

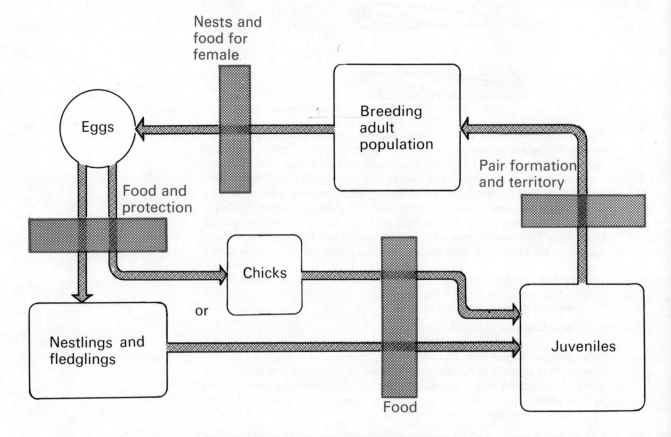

ITQ 46 Recall from the Introduction (Section 9.0) the views of Lack and of Wynne-Edwards about population regulation and state at which places in the model there should be regulation according to their theories.

Read the answer to ITQ 46.

Consider first the evidence available about egg production in birds.

ITQ 47 How does egg production vary in the three species studied in Unit 8?

Read the answer to ITQ 47.

These three species certainly do not lay a constant number of eggs in a clutch, but the variations in number can be associated with the nutritional level of the female bird before she lays, sometimes several months before she lays. Wynne-Edwards would argue that in so far as the nutritional state of the female can be related to territorial behaviour, and this involves epideictic displays, his theory of limiting population by birth control is supported; but the simpler hypothesis is that the food supply directly affects the female. In the tawny owls, the number of pairs breeding in the wood showed very little variation after it had levelled out at about thirty, yet the average clutch in those years varied considerably depending on whether the rodent population was high or low.

egg production

Lack takes an evolutionary view of clutch size and has put forward evidence that pairs of birds that produce more eggs than the average for that species usually rear fewer young; this implies that the average number is that which leads under average conditions to the maximum number of offspring surviving to become independent. Note that there may be differences from year to year; the average must be calculated over at least five years. In Wytham Wood, the average for tawny owls is two and for great tits it is nine, but in any year the majority of birds may lay more or fewer eggs than average. For many birds, the

clutch size

female does not brood the eggs until the last one has been laid, so all hatch at the same time and have an equal chance of being fed and surviving. When there is a food shortage, there is a high probability that all the nestlings will die. Birds of prey and some others have the habit of beginning to brood before the clutch is complete so the eggs hatch at different dates; the first chick to hatch is likely to get the first chance of feeding through the period between hatching and fledging, whereas the last hatched will get food for growth only if the food supply is abundant. The first strategy is ideal for birds that use a supply of food which is very abundant for a short period of time (e.g. caterpillars), whereas the second strategy gives the best chance of survival for at least one young if food is difficult to obtain.

TABLE 18   Great tit brood size (eggs that hatched) and survival of young as indicated by recoveries of marked juveniles at least three months old.* (Rearranged from Lack, 1966.)

					Brood size with most recoveries per brood in various years							
5	6	7	8	9	10	11	12	13	14	15	16
1951	1961	—	1955	1949	1952	(1953)	(1953)	1947	1948	1948	(1963)
			1957	1950	(1953)	1959	1959		(1958)	(1960)	
				1952	(1955)	1960	1960		1962		
				(1953)	1959		1963				
				1954	1961						
				1956							
				1958							
				1959							
				(1962)							

* Note that the same year may occur in different columns if survival was equal for two brood sizes; figures in brackets indicate that survival of that brood size was very close to maximum for that year. (Data for years 1947 to 1963.)

For great tits, it is clear from Table 18 that in some years brood sizes much greater or much smaller than the average were most successful in terms of the number of juveniles that survived; but broods of average size (9) were the most successful (or nearly so) in nine out of the seventeen years listed.

Consider now the evidence available about mortality outside the breeding season.

ITQ 48   Which were the key factors for the three bird species studied in Unit 8 and which 'mortalities' appeared to be density dependent?

Read the answer to ITQ 48.

The annual adult mortality is about 67 per cent for the red grouse, about 50 per cent for the great tit and for one-year-old tawny owls, and less than 20 per cent for older tawny owls. For all these species, the availability of food at some stage of the year may have a profound effect either on the number of young hatched or on the survival of juveniles.

All three bird species studied in Unit 8 establish territories: those of the tawny owl are occupied for the rest of the bird's life and provide food through the year and nesting holes; the great tit territories are occupied for the breeding season and most of the food for the young comes from the pair's territory; the red grouse occupy territory for longer than the great tit and territorial behaviour begins months before the eggs are laid. The significance of territorial behaviour in the lives of these birds is clearly rather different. For tawny owls, the number of territories in Wytham Wood remains rather constant (apart from the crash in numbers following the really disastrous winter of 1946 and the build-up after this). The recruitment of young owls to the breeding population balances the deaths of adults—but recall that the number of young hatched is far from constant. For great tits, the number of breeding pairs and, consequently, the size of territories depend on the density of the population just before the breeding season (after the winter mortality); hatching success is related inversely to the density of the breeding population. For red grouse, the size of territory depends partly on the level of aggression of the cocks and this is determined partly by the state of the food over a period of months before the hens lay their eggs. There is evidence that mean territory size can also vary by direct adjustment in autumn to a change in food supply during the summer.

territorial behaviour

Choosing three out of eight thousand species of birds might give a biased impression of population regulation in birds. Certainly there is a great variety among birds in their feeding habits, breeding habits and general life-style. Lack

(1966) discusses thirteen species of birds which had been studied for more than four years in a quantitative way and eleven other species for which less information was available; these twenty-four species covered a wide range of habits and of bird orders. Here we discuss two other contrasted types of bird populations.

Shearwaters and their allies, the albatrosses, illustrate some of the extreme specializations of sea birds. The Antipodean Sooty and Slender-billed shearwaters and the British Manx shearwater *Puffinus puffinus* range widely over the open sea, feeding on squids, small fish and large planktonic crustaceans; they nest in burrows, almost always on small islands where there are no mammalian predators. All shearwaters lay only one egg at a time; these are comparatively large (one-sixth of the body weight). The incubation time is fifty days or more and the parents change places every six days in the Manx shearwater (longer for other species). The feeding grounds may be a long distance from the nesting burrows. The chicks are usually brought large meals about every second day, and put on so much fat that they eventually weigh about half as much again as their parents. The parents desert the chicks which live for about two weeks on their fat, growing their feathers, and then leave for the sea; this movement is about 70 days after hatching for the Manx shearwater; the wandering albatross *Diomedea exulans* takes just over a year from the start of breeding to the chick being ready to leave the island.

Manx shearwater

The very slow rate of growth of shearwaters and albatrosses is presumably related to their food supply being sparse and variable. When nine pairs of shearwaters were each given an extra egg to brood, only three young shearwaters were reared out of a possible total of eighteen, compared with forty-two out of forty-four young reared by control pairs, each with only one egg. Young Manx shearwaters return to the breeding colonies when two years old, but most of them breed first at five years old.

ITQ 49 From the information given so far, what sort of mortality would you expect among the adult shearwaters?

Read the answer to ITQ 49.

Wood pigeons *Columba palambus* are considered pests by many farmers; they feed on grain and the leaves of clovers and brassicas (cabbages, etc.). They nest in small trees in hedges and copses and roost in winter in copses; their present habitat is very much man-made, resulting from farming practice. In south Cambridgeshire, Murton and his collaborators found that the birds bred between March and October, but with a peak in July to September at the time of the grain harvest, when food for the young was most plentiful; they could rear two, possibly even three broods annually. The birds need to spend the whole day feeding between November and spring to obtain enough food for maintenance. Pigeons lay two eggs and both parents incubate; if the eggs are left unattended, as happens if the food supply is sparse, they may be taken by predatory jays and magpies. Up to 80 per cent of the eggs may be taken by predators when food for the pigeons is sparse; the more closely spaced the nests, the more eggs are likely to be lost.

wood pigeon

Population numbers are highest at about the end of September and show a marked reduction by December and a further marked reduction by March after which the numbers begin to increase again. The proportion of juvenile birds in the population was about 60 per cent in September, but between 10 and 50 per cent in February; the highest loss of juveniles in early winter was in years when grain was scarce. Pigeons feed in flocks and there is a social hierarchy in which juveniles come lower than adults. In disputes over food, adults will win and there is evidence that juveniles take items which adults would reject. Mortality between December and spring could be correlated with the relative scarcity of the clover supply, so possibly the birds die of starvation. Average numbers were: 63 adults (in 40 ha) produced 91 fledged young (2.4 per pair), giving an autumn population of 154 which fell to 98 in December and 70 (49 adults, 21 juveniles) in February; when breeding started, there were 44 older birds and 19 yearlings (giving the 63 adults per 40 ha again).

social hierarchy

ITQ 50 What was the average mortality of adults during the year? What was the mortality of juveniles between autumn and the breeding season?

Read the answer to ITQ 50.

The main mortality in winter was density dependent and could be inversely related to the amount of food present. There was heavy shooting in winter, but the numbers killed were small in comparison with the number that disappeared (possibly through starvation). Murton concluded that shooting played no part in regulating the population. In birds that feed in flocks, such as pigeons and rooks, the existence of a social hierarchy is a mechanism by which juveniles and other 'low class' birds may be eliminated, while the 'upper class' birds get sufficient food and survive to breed in the next year. This is contest competition for the food supply.

In the long term, the amount of food available must set an upper limit to the bird population likely to be found in an area. Other factors may operate to regulate populations below the numbers set by food limitation, e.g. holes in tree trunks may limit numbers of birds that nest in such holes, such as great tits and spotted flycatchers.

Population numbers fluctuate and the discussion has centred on whether the numbers of the breeding stock in an area are directly limited by food, as Lack suggested, or whether the fluctuations are related to changes in food supply through the birds' social behaviour, as Wynne-Edwards proposed. The actual number present depends on natality, mortality and movements (immigration and emigration). For tawny owls, natality is directly related to food, but recuitment of juveniles to the adult population of Wytham Wood is limited by the stable territorial system; surplus juveniles emigrate or die. Great tits' breeding success is related to food supply and is density dependent, but the adult population is determined by 'survival' during the autumn and winter and there is not enough information to relate this to specific factors. In the case of the red grouse, the breeding stock is limited by territorial behaviour and not by food, but the territorial behaviour is adjusted to food supply and food influences territorial behaviour. The long-lived shearwaters have a very low natality rate and almost nothing is known about mortality factors. Wood pigeons' breeding success is related to food supply and the breeding stock is determined by winter survival which is partly the result of contest competition.

9.4 Population regulation in mammals

> **Study comment** Much of the discussion in this Section is about birth rates (reproductive strategies). You should refer to Sections 8.5.1, 8.5.2 and 8.5.3 for the basic information about five species of mammals. Again, note how far the observations seem to support the views of Lack or of Wynne-Edwards.

It is typical of mammal populations that the generations overlap. The mother suckles her young and there is for at least a short time a family group of mother and young perhaps with the father as well. Between the breeding seasons, the animals may be dispersed or the groups may persist or may form part of larger groups—herds or packs. When drawing up a life table, it is important to be able to age the individuals; this is often possible from observations on teeth or horns or bony parts, but it is sometimes difficult and the methods can usually be applied only to dead animals. For some species, the best that can be achieved is to classify the individuals into adults of the two sexes, juveniles, and young still dependent on their mother.

deducing age of mammals

Some of the other problems of studying mammal population dynamics were illustrated by the case studies in Unit 8, Section 8.5. Small, burrowing, nocturnal animals are difficult to observe and to trap; large animals that lurk in woodland are also difficult to count; and animals that migrate present problems in keeping them under observation but it may be possible to estimate numbers as they pass through an area. With large or scattered populations, there are problems of taking samples and extrapolating from these to the rest of the population.

sampling problems

The case studies were chosen as examples of contrasted types of populations: the voles are a classic example of herbivores showing fluctuations with a fairly regular periodicity; the wildebeest spend most of their lives as part of large herds which may migrate or may remain resident and whose numbers seem to be regulated in different ways in different areas; the moose on Isle Royale are a classic example of herbivores that overgrazed the habitat periodically, showing population fluctuations, until the wolves arrived and acted as regulating agents; the lions and the wolves are both territory-holding 'top' carnivores but appear to differ in recruitment strategy.

The population changes in mammals can be illustrated by a diagram as in Figure 19.

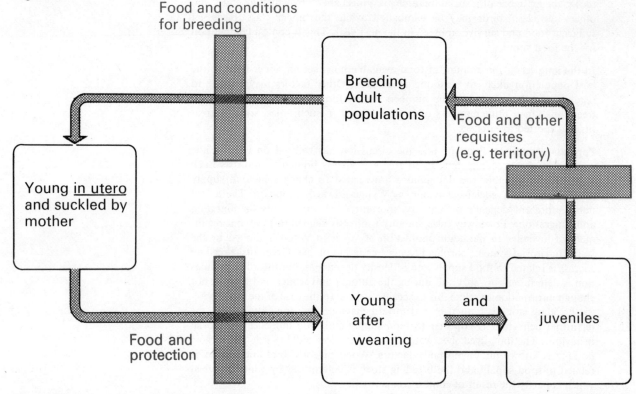

Consider first the evidence available about birth rate in mammals.

Figure 19 Model for mammal population regulation.

> **ITQ 51** All placental mammals are viviparous, but the numbers of young born simultaneously differ in different species. Try to suggest associations between number of young produced per pregnancy and the habits or habitats of the mammals.
>
> *Read the answer to ITQ 51.*

Generally speaking, the larger herbivores (e.g. deer, antelopes, elephants, rhinoceroses) produce a single or twin young per birth, whereas the majority of carnivores produce at least four young at a time in a den or lair. Rabbits and most of the rodents are small herbivores and these typically form nests or burrows and produce many helpless young at a birth; but hares and guinea pigs are exceptions in producing few well-developed young and not making nests. Primates (monkeys, apes, man) typically produce one or two young at a time, and the marine mammals (whales, dolphins, seals) typically produce a single young at a birth.

number of young

The average number of young produced and the frequency of breeding must bear a relationship to the pattern of mortality and relative length of life of animals of that species.

> **ITQ 52** Compare the reproductive potential of a female moose, which begins to breed in her third year and may produce one calf annually until she is fourteen years old, with that of a female wolf, which also begins to breed in her third year and may produce six pups annually until she is ten years old.
>
> **ITQ 53** Compare the reproductive potential of a female vole, which begins to breed when four weeks old and produces litters of four young every six weeks until she is 60 weeks old, with the female moose described in ITQ 52.
>
> *Read the answers to ITQs 52 and 53.*

The three species referred to in ITQs 52 and 53 clearly must have very different average potential birth rates: for the moose, 12 calves spread over 12 years; for the wolf, 48 pups spread over 8 years; for the vole, 36 young spread over

54 weeks. These are annual rates of 1, 6 and about 35 young per female. Another way of expressing the rates is in terms of 'generation time'—the interval between the breeding of the parents and the first breeding period of the young. For the moose, the generation time is three years (allowing for gestation time and annual breeding cycle), for the wolf also three years, and for the vole ten weeks (allowing for gestation time and assuming that breeding is not seasonal).

Dividing the total number of young that a female could produce by the generation time gives 4 moose calves, 16 wolf pups and 3.6 vole young. From this calculation, the values for the two herbivores are not very different, but the carnivore has a reproductive potential four times as high.

Consider whether the reproductive potential is actually fulfilled, i.e. whether the birth rate is relatively constant in all populations of the species or whether it is variable.

actual reproductive rate

> ITQ 54 From the information in Unit 8, quote examples to show that birth rates may vary within a species of mammal.

> *Read the answer to ITQ 54.*

For the three herbivore species studied in Unit 8, there seems to be evidence that the numbers of young produced vary, depending on the population density and on the way in which it has changed. The two carnivore species seem to have very different 'reproductive strategies': the wolves have a low birth rate, possibly related to their social structure, whereas the lions have a relatively high birth rate.

The evidence about birth rates suggests that the mammals studied produce at least twice as many young as are needed to replace the adults; this is based on the figure of four young per female per generation time for moose, voles and wolves. Consider the fate of these young.

mortality of young

> ITQ 55 From the information given in Unit 8, at what stages are there significant mortalities or reductions of numbers in the populations studied?

> *Read the answer to ITQ 55.*

Wolves and lions both form social groups and occupy territories within which they find their food and produce their young. Lions seem to produce relatively more young, but there is high mortality of cubs; the lionesses do most of the hunting and rear the cubs on a communal basis. The wolves produce fewer young since usually only one female in the pack breeds; the other wolves help to feed the cubs when they are old enough to take semi-solid food. Adults all cooperate in hunting and there is a strict hierarchy governing access to food.

Both these species are 'top carnivores'; on a global scale, their principal predator now is man. Both have been exterminated over much of their former ranges, but lions are now protected in Game Reserves, whereas wolves are still persecuted, except in quite small areas such as Isle Royale.

Wildebeest live in very large herds, whereas moose are solitary for most of their lives. The former feed by grazing on grasses and other herbs, and the moose feed mainly by browsing on trees, shrubs and aquatic plants (for which they can swim and dive).

World populations of both these large herbivores are much smaller now than one or two centuries ago. Wildebeest compete for grazing with cattle, but now have precedence and protection in game reserves; moose are hunted for sport, but probably still exist in small numbers over much of the suitable areas of North America.

Voles are too small and inconspicuous to be persecuted or preserved, but their well-being is probably much affected by agricultural practices; they are characteristically animals of grasslands or young woodlands.

There is not enough really detailed information about numbers for useful analyses to be made of 'key' factors and density dependent regulating factors for mammal populations, especially those of long-lived species but data are accumulating. (See SAQ 5, which you could now attempt if you have not already done so.)

There is further discussion of mammal population numbers in Sections 9.6 and 9.7.

9.5   Rates of increase of populations

Study comment In this short Section, the discussion about mammal birth rates is extended to other organisms. You could read through this quickly.

Populations may increase as a result of immigration exceeding the total of emigration plus deaths, but the most common mechanism of increase is through reproduction of the individuals resident in the population. The intrinsic rate of increase depends on the number of individuals capable of reproduction and on the number of new individuals (eggs or young) which each can produce in a given time.

ITQ 56 From information given in Unit 8, select examples of high and low intrinsic rates of increase.

Read the answer to ITQ 56.

It is necessary to distinguish between the maximum possible intrinsic rate of increase, which depends only on the fecundity of individuals and the size of the population, and the observed rate, which measures the actual reproductive performance. The discussion of 'potential birth rates' of wolves makes this distinction clear (Section 9.4). The maximum intrinsic rate of increase may be the only figure from which the starting numbers in a cohort can be reckoned; this is often true of fish populations. An estimate of average intrinsic rate of increase is used in calculations of mortality factors affecting bird populations (see Unit 8, Section 8.4).

maximum and actual rates of increase

The wolf population of Isle Royale is the only one of the case studies in Unit 8 for which there is some information about how the population increased after the first pair of wolves reached the island. The data from Table 16 of Unit 8 are plotted in Figure 20.

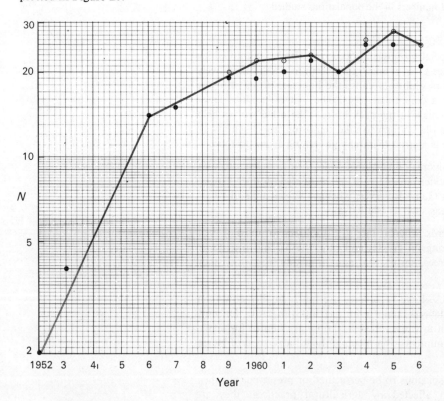

Figure 20 Population increase curve for the wolf population on Isle Royale: ● minimum number probably present; o best estimate of numbers present (including those found dead). (Data from Unit 8, Table 16.)

ITQ 57 Describe in words how the rate of population increase of wolves on Isle Royale has varied since 1952.

Read the answer to ITQ 57.

This curve is similar to the classical 'logistic curve' of population growth and reaches a steady level (an asymptote) of 24 wolves on the island. It can be interpreted as an example of a population which starts by reproducing freely and then suffers some restraint, as a result of which the rate of increase falls off proportionally as the numbers increase. The restraint in this case is presumably the area of territory available for the wolves, since it is unlikely that food supply is limiting.

When individuals of a species establish themselves in a new and favourable environment, there is commonly a rapid increase in population and the rate then falls off sometimes to a steady state of zero increase and sometimes to regular or irregular fluctuations below and above zero. Very high rates of increase may be shown by unicellular organisms which reproduce by dividing so that the population numbers double at frequent intervals. Populations in which the females reproduce parthenogenetically, such as cladocerans (e.g. *Daphnia*), rotifers and aphids, also may show extremely high rates of increase in the early stages of colonization of a new habitat or when conditions become favourable for growth (e.g. spring in temperate waters).

9.6 Herbivorous animals and their food plants

Study comment This is a general Section, covering all herbivores studied in Block B, including *IPE*, and attempting to extract generalizations from the data presented earlier.

Herbivores may be generalists, feeding on many species of plants, or specialists; on the whole, insect herbivores are more often specialists, feeding on one or a few species of plants only, than are mammalian or avian herbivores. Of course there are specialist mammalian herbivores, e.g. koalas (which eat leaves of certain *Eucalyptus* species only), and generalist insect herbivores, such as locusts. Generalists and specialists both specialize in the methods by which they consume their food, e.g. some bite and chew leaves, others suck plant juices, others eat fruit or bark.

The TV programme 'Introduction to Ecosystems' shows herbivores in grassland and woodland.

From Block A you may well have gained the impression that grazing mammals may have marked effects on grasslands, but herbivorous insects have relatively little effect on plant populations.

ITQ 58 In this Block, you have met evidence that specialist herbivorous insects can regulate populations of plants at low levels. Recall examples of this.

Read the answer to ITQ 58.

Locust 'plagues' illustrate what can happen when a generalist herbivorous insect undergoes a population explosion. You will recall also that 'defoliator' insects may reduce the net production (especially of wood) of forest trees. Other insect 'pests' of growing plants may affect their reproductive potential. The activities of these pests are obvious because their food plants are cultivated and often are concentrated as monocultures.

A much more subtle role for insect herbivores has been suggested—e.g. by Janzen (1970) and Connell (1970)—that in tropical forests the activities of specific herbivores result in the spacing out of their food tree species. The trees can support a large population of insects without being killed, but seedlings are not able to survive under insect attack, especially since they may be living under conditions that restrict growth, such as low light intensity on the forest floor. Thus seeds that fall from the parent tree and germinate close to it are likely to be killed by herbivorous insects; only if the seedling is sufficiently far away from other plants of its species not to be attacked by these insects has it a chance of survival. Mortality of 100 per cent of seedlings close to parent trees as a result of insect attack was recorded in experiments in Queensland, Australia.

The TV programme 'Tropical forest' shows seedlings under forest trees.

By preventing establishment of seedlings close to older trees of the same species, the specific herbivores maintain diversity in tropical forests; since each tree species is attacked by specific herbivorous insects, seedlings of other tree species can survive. Recall from Unit 5, Section 5.4.2, that high species diversity, which is associated with many specific herbivores and carnivores, is characteristic of mature ecosystems; the tropical forest is a very good example of this. The phenomena of diversity and various aspects of plant–insect relationships are discussed in Block C.

specialist herbivores in tropical forests

The activities of mammalian herbivores can affect the species present as well as the biomass and net production of the total vegetation. Recall the effects of the moose on the vegetation of Isle Royale—the ground hemlock *Taxus canadensis* over-grazing was exterminated and water lilies and pondweeds almost disappeared from the small lakes. Moose were able to reduce the level of their food supply to such an extent that many animals died of starvation and there are other examples of deer populations that have suffered the same fate, e.g. the deer of the Kaibab Plateau (see Unit 5, Section 5.2.3). Large herbivores in Africa have sometimes overgrazed their food supplies and caused deterioration of their habitats, e.g. elephants uprooting trees. These mammalian herbivores are typically preyed on by carnivores; in cases where severe over-grazing has occurred, the carnivores have often been scarce or absent (see next Section for further discussion).

Small herbivorous mammals such as voles and lemmings show cyclic fluctuations in numbers; this must mean cyclic changes in cropping pressure on their food plants. Investigations at Point Barrow have shown dramatic changes in vegetation at different stages of the lemming cycle. When voles are very common, the composition of the grasses probably changes and leaves of their favourite food plants may disappear. The underground parts of the plants may survive and be able to grow vigorously again when the vole population declines.

9.7 Carnivorous animals and their prey species

Study comment This Section attempts to extract generalizations from the data presented earlier about carnivores; in particular, there is discussion about whether vertebrate predators exert a regulating influence on the numbers of the herbivores on which they prey. Since *Homo sapiens* acts as an 'overlord carnivore', this discussion is very relevant to human activities.

Carnivores, using the term in its widest application, may be specialists or generalists.

ITQ 59 From information in this Block, quote examples of specialist carnivores.

Read the answer to ITQ 59.

There are generalist insect carnivores, such as the ground and rove beetles (Carabids and Staphylinids) and some that specialize in taking a restricted type of prey (as ladybird beetles (Coccinellids) take aphids (Homopterans)). Among the enormous taxonomic group of insects, more than 10 per cent are specialist carnivores and it is likely that the populations of many herbivorous insects are regulated by their specific predators and parasitoids. The principal reason why some insects are common and others are rare may be that the specific predators and parasites of the latter are more efficient in seeking out and consuming their prey than the specific predators of the common insects. specialist carnivorous insects

It is typical of carnivorous vertebrates that they are specialists in their methods of catching their prey. Compare the strategy of lurking and then darting, shown by pike, with the strategy of sitting in a tree and dropping down on rodent prey, shown by tawny owls; contrast these with the searching out of insect food by titmice and the group hunting techniques of wolf packs and lion prides. From your general knowledge, you can add many other specialist techniques of catching prey, e.g. the use of fangs and venom by snakes. carnivorous vertebrates

Few vertebrate predators are narrow specialists in the prey species taken, but many show preferences for some species rather than others.

ITQ 60 Quote examples of this from information in Units 8 and 9.

Read the answer to ITQ 60.

It is generally recognized that the activities of insect predators may regulate their prey numbers at densities far below those which the prey species would reach were the predators not present, but there has been much controversy about whether vertebrate predators ever act as regulating factors on the populations of their (usually vertebrate) prey. Errington (1946, 1963) has argued that verte-brate predators take only the 'doomed surplus' of their prey populations, mean-ing that the number of individuals they eat is equal to the number that would die anyway without reproducing because the population would otherwise exceed the 'carrying capacity' of the habitat. He thus implied that herbivore populations were capable of regulation of numbers in the total absence of the predators.

regulation of herbivore populations

> **ITQ 61** Quote evidence given in this Block that refutes this view.

> **ITQ 62** Quote evidence given in this Block of vertebrate herbivores that do not appear to be regulated by predation pressure.

Read the answers to ITQs 61 and 62.

Errington's hypothesis has been stated in slightly different terms by others: that predators take only the 'interest' and not the 'capital' components of the stocks of herbivores on which they feed. Some authors accept that the level of the 'capital' may have been reduced by the predation and this is virtually the same as accepting that the predators are, in fact, regulating the density of their prey.

Huffaker (1971) has reviewed 'predation', using the term in its widest sense to include predation by herbivores on their food plants. He quotes many examples with convincing evidence of regulation of vertebrate prey populations by their principal predators, including the wolves and moose of Isle Royale and the hyenas and wildebeest of Ngorongoro. He concedes that there are populations of vertebrate herbivores that are probably not regulated by predators, such as the migrating herds of ungulates on the Serengeti and some small rodent populations.

Huffaker discusses the interesting problem of the rarity or otherwise of predators in a given area and argues that 'controlling' predators must be present in sufficient numbers to take the 'interest' on the prey population's capital, i.e. the average loss to predators must balance the average annual increase in the prey population from reproduction. If a predator requires a high intake of its prey, the numbers of the predator must be low in relation to the numbers of prey. This is well illustrated by the wolves of Isle Royale:

density of predators

It appears that the stable population is about 24 wolves in the 544 km² of the island, giving an average density of 1 wolf per 23 km². The number of moose (using the figures of Jordan *et al.* 1971) is about 1000, giving a density of just over 40 moose for each wolf. The increase in moose population during each year is about 40 per cent and this is the number that must be cropped by the wolves or lost for other reasons if the moose population is to remain stable. Jordan *et al.* calculated that the wolves take about 124 adults and about 200 calves, a figure representing about 14 individuals per wolf, a crop of about 35 per cent of the stable population density of moose. Comparing other populations of moose and of wolves, Mech (1970) comments that the conditions in Isle Royale represent the highest number of prey per predator in a moose population which the wolves appear to regulate. Other populations with larger numbers of moose per wolf represent conditions in which the wolves do not regulate the prey numbers (usually the wolves are subject to predation by human hunters in these situations).

The 'overlord carnivores' studied in this Block—the tawny owls, lions and wolves —all hold territories in which they obtain most or all of their food. Huffaker believes that the evolution of the behaviour of maintaining territory (especially if territory size is to some extent compressible) is of advantage because it allows predators to crop a maximum yield of their prey. If the territory is too large, then the prey numbers can increase until its population goes 'out of control', as happens when the wolves (or lions) are persecuted while the prey is preserved. There may be lower limits set to the size of territories by behavioural mechanisms of the predators themselves; thus the Isle Royale wolves may have territories of the smallest acceptable size. Schaller (1972) considers that lions in Manyara Reserve, at a density of one per 2.6 km², are probably at the limit of compressibility of pride territories. In Wytham Wood, the adult tawny owl population seems to stabilize at 31 pairs, each pair with an average territory of 16×10^4 m².

predator territories

> **ITQ 63** Why should it be of advantage to predators to regulate the numbers of their principal prey?

Read the answer to ITQ 63.

Comparison of vertebrate and insect predators reveals one interesting difference: the numbers of breeding individuals of parasitic insects in balance with their hosts may be as high as five to one whereas, for vertebrates, the balance of predator to prey is closer to one to two hundred (for breeding female wolves on Isle Royale). This reflects differences in the intrinsic rate of increase of the prey, since the insects have a very high intrinsic rate of increase, but the ungulate populations produce less than one young per pair of the whole population. The insect predators need to crop a high rate of interest, whereas the vertebrates must adjust to a low rate of interest. The relative numbers of predator and prey must reflect these differences in their reproductive potential and that of their prey.

It is characteristic of predatory vertebrates that they switch to other food if their preferred prey is scarce. In any community, several species of vertebrate predator may coexist, feeding on the same 'pool' of herbivores; each species typically has a different preferred prey, but in times of scarcity they may all be subsisting on the same prey species. Generalist insect predators of different species also may coexist in communities, each with a preferred prey, but all capable of turning to other species which are temporarily abundant. The specialist insect predators are wholly dependent on one or a very small number of prey species, so their numbers are closely dependent on the specific prey and they must become extinct if it disappears completely from the habitat.

selection of prey

9.8 Mechanisms of population regulation

Study comment This Section summarizes the general principles developed in Block B.

Some species of organisms are always common in a particular ecosystem, others are always rare. Some species seem to be present in very similar numbers from year to year, whereas others vary in abundance, sometimes showing a thousand-fold change in three or four years. Some population fluctuations are very erratic, others are so regular that they are called cycles. All these are phenomena which should be covered by a general theory of population regulation if such a theory is possible.

In a closed population, with no immigration or emigration, any increase is the result of reproduction, and the potential rate of increase depends on the reproductive rate. Decrease is the result of deaths from whatever causes. In a stable population, reproduction is balanced by mortality; stability may be the result of adjustments in natality, mortality or both. When a closed population fluctuates, there must be variations in either natality, mortality, or both.

natality and mortality

The factors that affect natality and mortality can be classified as density independent, density dependent, inverse density dependent and delayed density dependent. Their effects on population numbers may be to disturb or to promote stability. Density independent factors may determine the level of population numbers at any time, but regulation of numbers at or about an equilibrium level requires some form of density dependence.

determination and regulation of numbers

Climate and physicochemical factors may limit the numbers of some species of organisms. One process by which population may be limited in a density dependent way is by competition (contest or scramble) for specific resources, such as refuges for flour beetles or nesting holes for some bird species. Some organisms may be rare because of limited tolerance or through scarcity of special requisites (this is discussed further in Block C).

competition

Weather factors are typically density independent and have been shown to determine population levels of many organisms, some of which show considerable fluctuations in numbers.

weather

Food supply may act as a density dependent factor regulating populations of some animals through competition; for others, it seems that the food supply is generally not limiting. In some cases a small component of the available food may be essential for the well-being of the animals. A limited food supply may regulate a population by reducing natality (e.g. tawny owls) or through increase in mortality (by starvation).

food supply

Predators may act as a density dependent or delayed density dependent factor regulating populations of their prey. In the widest sense, predator–prey interactions include herbivores eating their food plants and parasitic insects consuming their host insects, as well as carnivores feeding on varied types of prey. Some insects may be rare because they have efficient specific enemies. Successful biological control depends on identification, transfer and release of efficient specific enemies of pest insects or plants. Reduction in numbers of predators may result in catastrophic increase and fluctuations in numbers of their prey species (this is true of vertebrates as well as of insects).

predation

Behavioural responses of the individuals may lead to fluctuations or to stability of animal populations. These behavioural patterns include territorial behaviour, social hierarchy, emigration, immigration and 'stress' responses and may be related either directly or indirectly to level of food supply or to other environmental factors.

behaviour patterns

So many complex interactions between organisms and their environments are possible that it is not surprising that investigations are generally restricted to particular aspects of population changes or stability. Each restriction imposes limits on the types of analysis that are possible, and hence on the conclusions that can be reached. Hypotheses of population dynamics developed from different points of view for different species are often difficult to reconcile because relevant information is not available. There is great ignorance for many species about exactly when and how most individuals die. Semantic problems, such as the use of the word 'control' to include both determination of population level in one generation and regulation of population level over many generations, have led to misunderstandings and controversy. The situation is becoming clearer and it is apparent that the collection of bare statistics is not enough; the behaviour of the organisms must be studied and experiments set up (if possible) to test hypotheses. With detailed life tables, realistic models can be constructed and 'run' on computers to see how changes in natality or various mortalities could affect population levels.

restrictions of available data restrict types of analysis

None of the three views discussed in Section 9.0 gives a complete picture, but each presents certain facets of population dynamics and overlooks others. Much more information is needed before a general theory of population regulation can be stated with confidence.

9.9 Summary of Units 8 and 9

In evaluating any study of population numbers, it is necessary to consider how the organisms were counted; usually the counts are of samples, and the confidence limits of such samples must always be borne in mind.

Life tables are an important tool in the study of plant and vertebrate populations as also of insect populations. A 'horizontal' life table is based on a real cohort, whereas a 'vertical' life table is constructed from a census of the mixed ages within a population. Plots of survivorship curves can be of three main types, implying quite different patterns of mortality, as discussed in Section 8.1. The drawing up of life tables can lead to key factor analysis; this is illustrated with data for birds in Section 8.5.

life tables

Three main attitudes to the regulation of population numbers are outlined in Section 9.0. These are:

1 Andrewartha and Birch (1954) argued that changes in numbers could result from environmental pressures, such as weather or predators, and that numbers could be limited by shortages of resources and the inability of organisms to disperse in search of resources.. They believed that it was not necessary to postulate density dependent mortality factors.

2 Lack (1954, 1966) argued that populations fluctuate about medium levels and that density dependent factors must operate. He believed that reproductive rates vary much less than mortality rates and that the latter include density dependent elements which often operate through limitation of food supply outside the breeding season.

3   Wynne-Edwards (1962) argued that populations are ultimately limited by food supply, but rarely reach this limit because the individuals disperse; birth rates are restricted in response to 'epideictic' displays, and population numbers remain close to the 'optimum' for the particular environment.

The regulation of populations of flowering plants is discussed in Section 9.1 with particular reference to buttercups (described in Section 8.2.1). Stages at which regulation may occur are: when the seeds (from the 'reservoir in the soil') germinate; during subsequent growth, when the 'constant yield' principle operates; at the stage when seeds and/or vegetative propagules are produced. **flowering plants**

Population regulation in fishes is discussed in Section 9.2 with particular reference to perch and pike (described in Section 8.3.1) and brown trout (described in Section 8.3.2). Fish have a very high fecundity which is related to the size (and hence usually to the age) of the females; some lay very small eggs and there is generally a metamorphosis to the juvenile stage, others have large eggs and no metamorphosis. Brown trout acquire feeding territories as very young fry; contest competition at this stage regulates the population numbers, and mortality of older fish is usually at a low rate. Perch and pike populations are characterized by 'dominant year-classes', when survival of juveniles is high in contrast to other years when few juveniles survive. In Windermere, the pike numbers are probably now regulated by the netting carried out by the Freshwater Biological Association, and the perch populations may be regulated by predation by pike. **fishes**

The bird populations studied in detail are the tawny owls in Wytham Wood (Section 8.4.1), great tits in Wytham Wood (Section 8.4.2) and red grouse in in Scotland (Section 8.4.3); population regulation mechanisms are discussed in Section 9.3. For all three species, variation in egg production (number or quality) can be associated with nutritional level of the female before she lays. Adult tawny owl populations remain remarkably steady from year to year, but their egg production varies greatly and is related to food supply; population regulation seems to depend on movement of juveniles when they leave their parents' territory. Disappearance (possibly due to death from starvation) of juvenile great tits probably accounts for much of the key factor mortality, and the population seems to be regulated by the degree of success of the adults in producing hatched young. The quality of the eggs laid by the female red grouse seems to determine the level of aggression of cocks and hence their territory size, and this affects the level of nutrition of their mates. Birds such as albatrosses have a very low rate of reproduction and rate of growth; this is accompanied by a very low rate of mortality. Pigeons suffer density dependent mortality, probably from starvation in winter; as a result of hierarchical behaviour, it is mainly the juveniles that disappear. **birds**

Population regulation in mammals is discussed in Section 9.4; the populations studied in detail are voles (in Section 8.5.1), lions and wildebeest (in Section 8.5.2), and moose and wolves (in Section 8.5.3). The vole populations are subject to fluctuations related to high death rates after peak populations, but the reasons are not clear and are probably different in different places. The moose population on Isle Royale is now regulated by the wolves; wildebeest populations are regulated by predators in some areas, but probably malnutrition and disease are the major mortality factors in other areas. Lions and wolves both form social groups that hold territories. The lion populations seem to be regulated by emigration of young males; there is a comparatively high reproductive rate and high mortality of cubs even when food is not limiting. In wolf packs, breeding is restricted to the alpha female and cub mortality is lower than for lions. **mammals**

Fish populations have potentially a high rate of increase, whereas moose and wildebeest have a low potential rate of increase. The wolf population on Isle Royale seems to have increased along a 'logistic' curve and is now presumably the largest number that can fit into the area available as territory.

Although, in general, herbivores take only a small proportion of plant production, there are examples of herbivores regulating plant populations at low levels (e.g. *Opuntia*, regulated by *Cactoblastis*) and of herbivores severely overgrazing plants (e.g. locusts, moose before the wolves came). Specialist herbivores may play a subtle role in maintaining diversity in tropical forests.

Specialist insect carnivores (e.g. parasitic insects) can regulate herbivore populations at low levels. Evidence that vertebrate predators can act as regulating factors on their vertebrate prey is quoted in Section 9.7, refuting the hypothesis that they always take only a 'doomed surplus'. Insects 'crop' prey with a high intrinsic rate of increase, whereas terrestrial vertebrates must adjust to prey with a low intrinsic rate of increase. Several species of generalist predators may coexist, sharing a 'pool' of herbivores, but each with a different preferred prey.

The general principles discussed in Block B are summarized in Section 9.8.

Publications cited in Units 8 and 9

Allen, K. R. (1951) The Horokiwi Stream—A study of a trout population, *N.Z. Marine Department Fisheries Bulletin No. 10*, Wellington, N.Z.

Andrewartha, H. G. and Birch, L. C. (1954) *The Distribution and Abundance of Animals*, Chicago University Press.

Connell, J. H. (1971) On the role of natural enemies in preventing competitive exclusion in some marine animals and in rain forest trees, in den Boer, P. J. and Gradwell, G. R. (ed.), *Dynamics of Populations*, pp. 293–312, Centre for Agricultural Publishing and Documentation, Wageningen.

Deevey, E. S. jr. (1947) Life tables for natural populations, *Quart. Rev. Biol.*, **22**, 283–314.

Elton, C. S. (1924) Periodic fluctuations in the number of animals: their causes and effects, *J. Exp. Biol.*, **2**, 119–63.

Errington, P. L. (1946) Predation and vertebrate populations, *Quart. Rev. Biol.*, **21**, 144–77, 221–45.

Errington, P. L. (1963) The phenomenon of predation, *Amer. Sci.*, **51**, 180–92.

Foster, J. and Kearney, D. (1967) Nairobi National Park Game Census 1966, *E. Afr. Wildlife J.*, **5**, 112–20.

Frost, W. E. and Smyly, W. J. P. (1952) The brown trout of a moorland fishpond, *J. Anim. Ecol.*, **21**, 62–86.

Frost, W. E. and Brown, M. E. (1970) *The Trout*, Fontana New Naturalist.

Harper, J. L. and White, J. (1971) The dynamics of plant populations, in den Boer, P. J. and Gradwell, G. R. (ed.), *Dynamics of Populations*, pp. 41–63, Wageningen.

Huffaker, C. B. (1971) The phenomenon of predation and its roles in nature, in den Boer, P. J. and Gradwell, G. R. (ed.), *Dynamics of Populations*, pp. 327–43, Wageningen.

Janzen, D. H. (1970) The unexploited tropics, *Bull. Ecol. Soc. Am.*, **51**, 4–7.

Jenkins, D. and Watson, A. (1970) Population control in red grouse and rock ptarmigan in Scotland, *Finnish Game Research*, **30**, 121–41.

Jordan, P. A., Botkin, D. B. and Wolfe, M. L. (1971) Biomass dynamics in a moose population, *Ecology*, **51**, 147–52.

Kalleberg, H. (1958) Observations in a stream tank of territoriality and competition in juvenile salmon and trout (*Salmo salar*, L and *Salmo trutta*, L), *Rep. Inst. Freshwat. Res. Drottningholm*, **39**, 55–98.

Kipling, C. and Frost, W. E. (1969) Variations in the fecundity of pike *Esox lucius* L. in Windermere, *J. Fish. Biol.*, **1**, 221–37.

Kipling, C. and Frost, W. E. (1970) A study of the mortality, population numbers, year-class strengths, production and food consumption of pike *Esox lucius*, in Windermere from 1944 to 1962, *J. Anim. Ecol.*, **39**, 115–57.

Krebs, C. J. (1971) Genetic and behavioural studies on fluctuating vole populations, in den Boer, P. J. and Gradwell, G. R. (ed.), *Dynamics of Populations*, pp. 243–56, Wageningen.

Krebs, J. R. (1970) Regulation of numbers in the Great tit (Aves: Passeriformes), *J. Zool.*, **162**, 317–33.

Krebs, J. R. (1971) Territory and breeding density in the great tit *Parus major* L, *Ecology*, **52**, 2–22.

Kruuk, H. (1970) Interactions between populations of spotted hyenas (*Crocuta crocuta* Erxleben) and their prey species, in Watson, A. (ed.) *Animal populations in relation to their food resources*, pp. 359–74, Blackwell Scientific Publications.

Lack, D. L. (1954) *The natural regulation of animal numbers*, Clarendon Press, Oxford.

Lack, D. L. (1966) *Population studies of birds*, Clarendon Press, Oxford.

Le Cren, E. D. (1955) Year to year variations in the year-class strength of *Perca fluviatilis*, *Verh. int. Verein. theor angew. Limnol.*, **12**, 187–92.

Le Cren, E. D. (1958) Observations on the growth of perch (*Perca fluviatilis*) over twenty-two years with special reference to the effects of temperature and changes in population density, *J. Anim. Ecol.*, **27**, 298–334.

Le Cren, E. D. (1965) Some factors regulating the size of populations of freshwater fish, *Mitt. int. Verein. theor. angew. Limnol.*, **13**, 88–105.

Le Cren, E. D., Kipling, C. and McCormack J. C. (1972) Windermere: Effects of exploitation and eutrophication on the salmonid community, *J. Fish. Res. Bd Can.*, **29**, 819–32.

Mech, L. D. (1966) The wolves of Isle Royale, *Fauna natn. Pks U.S.*, **7**, 210 pp.

Mech, L. D. (1970) *The Wolf: The ecology and behaviour of an endangered species*, Natural History Press, N.Y.

Murie, A. (1934) The wolves of Isle Royale, *Misc. Publs Mus. Zool. Univ. Mich.*, **25**, 44 pp.

Murton, R. K. (1965) *The Wood Pigeon*, Collins New Naturalist.

Myers, J. H. and Krebs, C. J. (1971) Genetic, behavioural, and reproductive attributes of dispersing field voles, *Microtus pennsylvanicus* and *Microtus ochrogaster*, *Ecol. Monogr.*, **41**, 53–78.

Perrins, C. M. (1965) Population fluctuations and clutch size in the Great Tit (*Parus major*), *J. Anim. Ecol.*, **34**, 601–47.

Sarukhán, J. (1971) *Studies on Plant Demography*, Ph.D. Thesis, University of Wales.

Schaller, G. B. (1972) *The Serengeti Lion. A study of predator-prey relations*, University of Chicago Press.

Southern, H. N. (1970) The natural control of a population of tawny owls (*Strix aluco*), *J. Zool.*, **162**, 197–285.

Spinage, C. A. (1972) African ungulate life tables, *Ecology*, **53**, 645–52.

Watson, A. (1971) Key factor analysis, density dependence and population limitation in red grouse, in den Boer, P. J. and Gradwell, G. R. (ed.), *Dynamics of Populations*, pp. 548–64. Wageningen.

Wynne-Edwards, V. C. (1962) *Animal Dispersion in relation to Social Behaviour*, Oliver and Boyd.

TABLE 20 Data for buffaloes in the Serengeti—(all $\times 10^3$).

	1965	1967	1969	1970
Population level after calving if all mature females had produced calves (T)	51	59	67	70
Actual population after calving (C)	47	55	61	64
Population of animals more than one year old (after juvenile mortality) (A)	41	44	51	51
Population of adults just before next breeding season (S)	37	41	45	45

Self-assessment questions

SAQ 1 (*Objectives 2* (*a*)) Plot the data for male and female buffaloes (Table 1, p. 6) against the age at death of the individuals expressed: (a) as percentage deviation from the mean length of life and (b) as a percentage of the total life span (19 years for females and 23 years for males). You could use Figure 1 (the graph paper provided for plotting the survivorship curves against the actual age at death), but make these curves in different colours.

SAQ 2 (*Objectives 1, 2(a), 3*) Identify which of the descriptions (a) to (e) fit the types of cohort survivorship data and curves (1) to (5).
(a) The mortality rate is approximately the same at all ages.
(b) All the individuals were born in the same year.
(c) Young individuals are subject to heavy mortality but older individuals have a high chance of survival.
(d) Most individuals survive well and all die at approximately the same age.
(e) Individuals caught in the same locality during one year were 'aged' by study of bones.

(1) positively skew rectangular   (2) negatively skew rectangular
(3) diagonal   (4) horizontal   (5) vertical

SAQ 3 (*Objective 6*) Identify which of the following concepts (A to L) are basic to the theories of population regulation or determination of (1) Andrewartha and Birch (2) Lack (3) Wynne-Edwards.

A   density dependent mortality from lack of food of juveniles or adults
B   very high intrinsic rate of increase
C   contest competiton for some requisite
D   inverse density dependent regulation of natality
E   density independent mortality, often due to weather
F   parental protection and feeding of young
G   relatively constant natality at the rate giving maximum survival of young under average conditions
H   homeostatic regulation of numbers
J   no density dependent regulation of numbers
K   display behaviour indicating relative abundance of individuals
L   territorial behaviour

SAQ 4 (*Objectives 2(a), 4(d), 5*) Sockeye salmon *Oncorhynchus nerka* spawn in North American rivers flowing into the Pacific Ocean; the fry live in lakes for nearly two years, then turn into smolts and migrate down to the sea. They spend nearly two years at sea, returning to spawn in their native stream four years after their parents spawned there. This life cycle is very regular in many Canadian rivers; there is often a regular pattern of variation in the numbers of individuals in the four year-classes as illustrated in Table 19 (where the figures have been simplified). In subsequent years, this pattern of numbers would repeat regularly so that figures for stages of the life history in year V and year IX would be within ten per cent of year I, for year VI and year X within ten per cent of year II and so on.

Examine Table 19: note that one of the four year-classes is markedly different from the other three. Try to decide how to analyse the figures in an attempt to explain the difference, then continue with this SAQ.

(a) Calculate the k-values for each interval in the life history and K for each year-class. Describe briefly how the k-values for different stages in the life history differ. Then describe how K and the successive k-values vary for different year-classes.

Now suggest an explanation for the differences in numbers between year-classes.

(b) Calculate the apparent fecundity of the salmon of different year-classes (assuming that all adult fish spawn and the sex ratio is 1:1).

(c) A large number of the adult salmon are caught by fishermen *before* they spawn. Assuming that the actual fecundity is the same for every year-class, what proportion of salmon of year-class II are caught by fishermen compared with the proportion caught of year-class I?

SAQ 5 (*Objectives 2(a), (b), (c), 5*) Observations on buffalo *Syncerus caffer* in the Serengeti have recently been used to produce models to investigate population regulation (Sinclair, A. R. E. (1973) 'Regulation and population models for a tropical ruminant', *E. Afr. Wildl. J.* volume 11 (in press)). The figures in Table 20 (p. 60) have been simplified for the purpose of this exercise and represent only four out of the eight years analysed.

(a) Are these life tables and, if so, of what type?

(b) Calculate the k-values for: 'loss' due to failure to calve (k_1); juvenile mortality (k_2); adult mortality (k_3); total annual mortality (including 'loss' due to failure to breed) K.

(c) From about 1969 on, the population of breeding adults remained fairly constant; in 1965 and 1967, the population was increasing. From the k-values, try to suggest changes in mortality that could account for these population changes.

(d) A model was set up assuming that k_3 is density dependent, and using constant average values for k_1 and k_2. Starting with the population figures for 1965, the predicted population levels closely followed the observed population levels and predicted an equilibrium population of about 56×10^3 animals over one year old. (The actual number in 1972 was 53×10^3 adults plus yearlings.) What do you deduce from this information about the processes regulating the population of buffaloes?

(e) The same model was used to extrapolate backwards from 1965 to 1958. The curve was unlike the real data and gave population levels very much lower than those observed. What do you deduce from this information?

(f) A second model was set up assuming that $k_2 + k_3$ is density dependent with k_1 as a constant. This model predicted values for 1965 and later, which were lower than those recorded; it predicted an equilibrium population of 48×10^3 (exceeded by 1969). It also predicted values before 1965 lower than those recorded and even lower than those obtained with the first model. What do you deduce from this information?

(g) Buffaloes are very susceptible to rinderpest, a disease of which there were epidemics in Serengeti in the 1950s; this disease disappeared in 1962–3. In a wildebeest population, rinderpest killed all susceptible adults early in the outbreak and in subsequent years the disease killed calves, resulting in wildebeest calf mortality 1.76 times as high as when there was no disease. Suggest how this information could be used to investigate buffalo populations before 1965.

TABLE 19   Data for sockeye salmon in a Canadian river.

Year class	I	II	III	IV
Number of eggs (E)	15×10^5	32×10^7	75×10^5	60×10^5
Number of fry (F)	3×10^5	16×10^7	15×10^5	12×10^5
Number of smolts (S)	3×10^4	8×10^7	15×10^4	13×10^4
Number of adult salmon (A)	1×10^3	2×10^6	5×10^3	4×10^3

ITQ answers

ITQ 1 The horizontal life table deals with an actual group of individuals whose mortality rates will to some extent be governed by events in the ecosystem at each age interval. If there is a drought when they are nine years old, the mortality for that age interval may be very much higher than that for age eight or age ten. The cohort belonging to the following year or to four years later will have a different pattern of mortality and will show higher than normal mortality in their eighth and fifth year respectively. The vertical life table summarizes the mortalities affecting all the age groups that are living at the same time in a population. If the study is based on data for different age groups collected over a long period of time, as when skulls are collected in an area and their ages at death then estimated, the life table that is constructed may be a very reliable average but individual year classes may have suffered mortalities quite different from the average.

ITQ 2 When numbers of survivors are plotted on a logarithmic scale against time on an arithmetic scale, the slope of the line between the points shows the rate of change of population size—in this case, the mortality rate. In the completed Figure 1, straight lines have been drawn to show stages in the life cycle when mortality rate stays constant or changes. For both sexes, mortality is high in the first year. For males, mortality rate is then low (the line is almost horizontal) up to the ninth year, when the slope of the line changes and mortality continues at a greater but fairly constant rate until the cohort is dead. The female survivorship curve is more complex and seems to have three periods with progressively higher rates of mortality, i.e. from one to five years, from five to eleven years and from eleven years to the end of the cohort's life at about eighteen years.

ITQ 3 Curve A represents the survivorship of a cohort in which there are very few deaths until the animals reach a certain age; at this age, the mortality rate suddenly increases and reaches 100 per cent. All the animals die within a short period of time. This is called a 'negatively skew rectangular' curve.

Curve B represents a cohort in which the rate of mortality remains constant throughout life. This is called a 'diagonal' survivorship curve.

Curve C represents a cohort in which the mortality rate is very high early in life, so that most individuals die soon after birth; those that survive have a constant low rate of mortality, so that a few individuals may live to great (comparatively) ages. This is a 'positively skew rectangular' curve.

ITQ 4 Mean life span can be calculated with fair accuracy from a limited number of observations, whereas estimating total life span depends on finding long-lived individuals. If the shape of the curve near the end of life is negatively skew rectangular, it is unlikely that the 'spread' of the curve will be altered much by making more observations, but for curves of the positively skew rectangular type, there is the possibility of discovering one or a few very long-lived individuals and this could greatly alter the position and shape of the curve. So Figure 3 is less likely to change shape than Figure 4.

ITQ 5

	R. acris	R. bulbosus	R. repens
(c)	+62	−26	+22
(d)	2	0.7	1.2
(h)	55	45.5	11
(k)	76	63	62

ITQ 6 The survivorship curves are shown in Figure 21. The seedlings of *R. acris* represent a real cohort and reveal a positively skew rectangular type of curve with high mortality early in life and then a much lower rate of mortality increasing between weeks 50 and 60 and then falling again. The other plants could be of mixed ages, but probably those of *R. repens* were two real cohorts mixed; these give a curve closest to the diagonal type with a fairly high rate of mortality, but there is a period of no mortality between weeks 80 and 105 and then death of all individuals within fifteen weeks. The curves for *R. acris* and *R. bulbosus* mature plants are similar to each other, but at different levels of population. Both are closest to the diagonal type with low rates of mortality through most of the study period.

ITQ 7 The relative increase in the population shows an inverse relationship to the number of plants m^{-2}. This is equivalent to density dependent mortality and could be part of the regulating mechanism for populations of this species.

ITQ 8 In each case, the greatest number of living seeds in the soil were those of the predominant species on the surface, suggesting that distribution is over short distances only. Seeds of all three species were found in all three plots used for this Table, but this observation did not apply to all the other plots examined.

ITQ 9 (a) There is high mortality:
For *R. acris*, between April and June after the seeds germinate; if all mortality is of seedlings, then 70 per cent of the seedlings die within 8 weeks.

For *R. bulbosus*, between October and December after the seeds germinate; more than 40 per cent of the seedlings may die in this period and another 20 per cent could die between December and April.

For *R. repens*, there are high mortalities between April and June and it is possible that all the seedlings and some mature plants die then; there is a high mortality between August and October and this probably represents death of mature plants that have produced daughters (more than 30 per cent of the mature plants die at this time).

(b) For *R. acris*, the highest number die between August and October, soon after the seeds have reached the soil; a high proportion of the surviving seeds die between April and June, the germination period (more than 40 per cent die and about 35 per cent establish themselves as seedlings).

For *R. bulbosus*, the highest number die between August and October, just before the germination period; these could be new seeds or old seeds.

For *R. repens*, the highest numbers die between April and June and June and August; since only 84 new seeds are added, most of these 219 deaths must be seeds of earlier years.

(c) *R. bulbosus* does not reproduce vegetatively; all new individuals are from seeds which must spend at least one summer in the soil before germination. Seed production in this model is about 12 per plant, and seedlings in April could represent half the population or more (assuming all deaths are seedlings, 61 had died and 34 had survived at this time).

R. acris produced 7 daughters at a time when the population was 72 (increase of 10 per cent); of this 72, at least 15 (20 per cent) must have been seedlings (assuming no deaths of mature plants, which is unlikely). Seed production was about 20 per plant and these seeds must spend at least nine months in the soil during which time there is probably high mortality.

R. repens produced few seeds, and it seems likely that none of the small number of seedlings survived beyond a few weeks. The number of daughters produced in October more than doubled the population at that time; this species relies on vegetative reproduction for population increase.

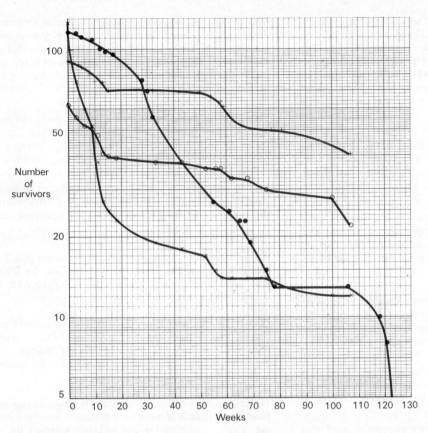

Figure 21 Survivorship curves for buttercups, based on data in Table 3. × seedlings of *Ranunculus acris*; ○ mature plants of *R. acris*; + all plants of *R. bulbosus*; ● all plants of *R. repens*.

ITQ 10 It seems that the growth of the older perch (in their third year and onwards) was restricted when they were more crowded, but that this restriction did not affect the younger fish. The rate of growth of fish depends, among other things, on food supply. The food supply of larger perch might be such that there is not enough for unlimited feeding by large numbers, so that quantity of food acts in a density dependent way. For smaller perch, the food might be completely different and always be sufficient not to limit growth rate.

ITQ 11 The decline in numbers of three-year-old fish in the year-classes 1941 to 1945 is associated with a steep fall in the numbers of eggs laid—presumably the result of the lowering of the numbers and average size of the adult population by the intense fishing. The percentage survival of the young fish varies between about 0.005 and 0.01 per cent for those years; it is higher in 1943 to 1945 than in the earlier years.

ITQ 12 No. The numbers of three-year-old perch of year-classes 1946 to 1950 were low for all except 1949 when the number was very much higher (about thirty times the 1948 year-class). The numbers of eggs laid varied from year to year and was about four times as many in 1947 as in 1949. The highest number of three-year-olds actually came from the lowest number of eggs produced, and it is evident that there were great variations in the percentage survival of the young fish and that it was this, and not the number of eggs laid, that determined the numbers of adults three years later.

ITQ 13 The synchronous variation in 'strength' of year-classes suggests that some climatic factor could be affecting young perch in all the lakes, e.g. the water temperature in June and July may affect the numbers of small zooplankton available as food for the newly hatched perch fry and so determine whether survival is good or poor.

ITQ 14 Both fish populations showed initial fall in numbers as the older, larger fish were removed. The perch populations have fluctuated about a new, lower level even though they have not been fished intensively since 1948 (North Basin) or 1964 (South Basin). The pike populations have been fished intensively every year yet the numbers have returned to the level of 1944 (and actually fluctuate about 50 per cent above that level).

ITQ 15 These fish are eaten by pike in autumn (char) and all through the year (trout), but they are usually sufficiently large to be taken by larger pike rather than smaller ones. The removal of larger pike should therefore result in better survival of the larger, older char and trout. Probably this has happened for the char and possibly for the trout, but there is much less information about these species than about the pike and perch.

ITQ 16 Mortality as a percentage:

between egg production and start of feeding	45
during three months after start of feeding	98
during next three months	37.5
during next three months	40
during second year of life	80
during third year of life	80
during fourth year of life	80
during fifth year of life	80

ITQ 17 The 21 000 eggs gave rise to about 130 three-year-old trout—a survival of 0.6 per cent, which is better than the 0.02 per cent survival in the Horokiwi. For fish more than three years old, the average survival in Three Dubs Tarn is about 35 per cent, which again is better than the 20 per cent for the Horokiwi River. Both waters show very high mortality in early life, but there is not enough evidence to establish when this happens for the Tarn population.

ITQ 18 Survival to the end of the first year is 2.5 per cent, much higher than the 0.5 per cent survival in the Horokiwi. After that, survival in the Lake District is 50 per cent (until anglers start to crop the stock), which again is much better than in the Horokiwi River.

ITQ 19 Above a limiting density of about 40 fry m^{-2}, there is a direct correlation between mortality and population density. Compare this Figure with 2.11 of *Insect Population Ecology*. The survival of the trout is a good example of contest competition. All the fry up to the limiting density have a high probability of surviving, but increase above this density results in no change in the *number* of survivors, which means that the *percentage* surviving decreases.

ITQ 20 If the winter food supply is scarce, the female owls may be unable to obtain enough food to lay eggs; under these conditions you might also expect a higher level of desertion of eggs by hungry females (and so a higher k_3).

ITQ 21 k_5 is strongly density dependent; it is almost exactly proportional to the number of owlets that are fledged. This means that the greater the number of young owlets that are reared in the year, the greater the number that disappear (emigrate or die).

ITQ 22 k_4 varies in a very similar way to K (the total mortality) so k_4 is the 'key factor'. It is difficult to say whether or not k_1 or k_2 plus k_3 are density dependent and vary in such a way as to restore the population to a medium level.

ITQ 23 k_1 and k_2 are both significantly density dependent ($p < 0.01$); k_3 is not density dependent (there is no linear relationship between mortality and log population density).

ITQ 24 Juveniles. The ratio of adults to juveniles changed from 1:3 to 1:1 so many more juveniles died than adults.

ITQ 25 A decline in amount and nutrient content of heather in winter and spring (due to weather and the effects of grazing animals) results in poor maternal nutrition, poor egg quality and a small proportion of young reared. Later in autumn, the young cocks that take up territories are more aggressive and take larger territories than older cocks; this means an increase in mean territory size and consequently a decrease in the number of grouse breeding next spring.

An increase in breeding stock results from an increase in the amount and nutrient content of the heather, acting via good egg quality, large broods and young cocks being less aggressive and taking up smaller territories than those of older cocks. Some alteration of territory size in adjustment to food supply in later summer and autumn can also occur.

ITQ 26 The natural population rose to a peak of about eighty animals per trapping period in the autumn of 1967, fell during the winter, rose again the next spring, fluctuated and reached a peak of about sixty individuals, and then fell to less than two individuals in the spring of 1969. The fenced population rose to a winter 1968 peak of about sixty individuals, fell and then rose steadily to reach a winter 1969 peak of about three hundred individuals per trapping period.

ITQ 27 Krebs' group's work seems to indicate that failure of food supply does not usually occur, but there could be subtle changes in food supply if some species of plants are over-grazed. Soil nutrients might affect voles through the plants forming the food supply so the plants should be investigated first. The regularity of some cycles makes it unlikely that weather factors determine the cycles, although there may be disastrous years due to very bad weather. Observed changes in the average weight of dispersing animals and in their behaviour at different phases of the cycle suggests that physiological and behavioural changes related to population density are a very promising field of enquiry. Changes in genetic constitution have been documented, but it is not clear whether these are the cause or the result of the cycle so this needs further investigation.

ITQ 28 The individuals are spaced out as groups. 'Resources' are presumably spaced out and the obvious resource to consider is food supply. Living their whole lives in the same area allows the lionesses to have a very detailed knowledge of its topography including water holes and suitable ambushes; hunting is thus likely to require less expenditure of energy and be more efficient.

ITQ 29 The young male lions leave the pride and become nomads; some females also become nomads. Thus dispersal is the mechanism that operates, with mortality, to regulate the lion population at a fairly stable level over long periods of time.

ITQ 30 Multiply the prey biomass (expressed as kg km^{-2}) by the average area per lion (km^2) and divide by 2500 (kg) to obtain the reciprocal of the proportion of prey consumed, i.e. for Ngorongoro, $\frac{10\,363 \times 3.7}{2500} = 15$, so the lion consumes one-fifteenth of the prey biomass available per average lion. For the Manyara reserve, the proportion is one-eighth; for Serengeti, one-twentieth; for Nairobi, one-sixth; for Kruger (in South Africa) one-seventh.

ITQ 31 The biomass of spotted hyenas in Ngorongoro exceeds that of lions very considerably. In the other four reserves, the lions are the predators with the greatest biomass, but in Ngorongoro, lions are second to spotted hyenas (of which there are about 480 km^{-2}).

ITQ 32 Herds subjected to greater predation display greater fecundity as a result of more cows starting to breed at a younger age. This can be turned round to imply a density dependent mechanism working through fecundity, leading to a greater production of calves when the density of wildebeest is lower; the Serengeti figures fit this hypothesis.

ITQ 33 Lions selected wildebeest in preference to other animals, taking a higher proportion as the wildebeest numbers fell and causing an inverse density dependent mortality. When wildebeest became scarce, the lions turned to other more abundant species such as eland.

ITQ 34 (a) 441.6 calves give 126 (twice 63) yearlings, so the mortality is 316 and the mortality rate is 72 per cent.

(b) The mortality is 18 (twice 63 minus twice 54) and the mortality rate is 14 per cent. Both answers assume equal numbers of males and females.

ITQ 35 The higher twinning rate implies a higher rate of reproduction in 1959. The lower rate of reproduction in 1929 (when the population may have been 3000 moose) suggests that there may be a density dependent effect with higher twinning rate (higher fecundity) in the less-crowded, probably healthier, population. Compare this with the wildebeest cows (see the answer to ITQ 32).

ITQ 36 If about 10 per cent of the population of adults die, this is 2.4 wolves, and the recruitment should be between two and three yearlings in each year.

ITQ 37 The three females would produce 18 young. Suppose that 33 per cent survive to one year old—this gives six yearlings; if 50 per cent of these survive for another year, this is three two-year-olds. Thus the population in spring might consist of: 6 yearlings + 3 two-year-olds + 15 older adults. If the oldest adult is ten years old, a possible distribution of numbers is: three wolves of three years old, two wolves each of four, five, six, seven and eight years old and one wolf each of nine and ten years old (giving a total of 15). Your life table may be slightly different from this depending on what mortality/survival values you decided to assume. Each pack could consist of the dominant male and female plus two yearling pups, one two-year-old juvenile and three other adults, of whom one might be senile.

ITQ 38 The amount of water in the soil could act as a sieve; *R. repens* seeds can germinate in waterlogged soils, *R. bulbosus* only in soils that are well-drained.

ITQ 39 The seeds at point P were sown at low density, those at point Q were sown at high density. When harvested at 84 days, the P plants are each about ten times heavier than the Q plants. The total weight of plants produced on each plot was very similar in this experiment.

ITQ 40 This is a density dependent type of relationship; it will tend to regulate seed production at an intermediate level. If all the seeds develop at once, the next generation after a crowded generation will have fewer numbers, whereas the next generation after an uncrowded generation will have greater numbers.

ITQ 41 The seed reservoir, which persists sometimes over many years. Most animals produce eggs that hatch within a short time of their production. You should note, however, that some species of animals have eggs that resemble the seeds of plants in being able to remain dormant for an indeterminate number of years. The 'winter eggs' of zooplankton organisms such as rotifers and cladocerans (e.g. ephippia of *Daphnia*) fall into this category.

ITQ 42 According to Wynne-Edwards, the fish should regulate the number of eggs produced, depending on the density of the adult population. Lack believes that the animals lay the number of eggs that are likely to produce the largest number of juveniles; he considers that there is likely to be density dependent mortality out of the breeding season and that this regulates the population numbers. Wynne-Edwards believes in contest competition at some stage of the life history limiting the population so that the food supply is sufficient. Thus, according to Wynne-Edwards, population regulation should occur when mature fish populations produce their eggs, whereas Lack would expect it to occur at some other time.

ITQ 43 For the perch and trout, the number of eggs produced depends on the numbers in the spawning population and also on their sizes (since larger fish produce more eggs). There is no indication of population regulation at this stage. Shortage of substrate suitable for spawning could regulate the number of young produced, and this may happen for salmonid fishes which spawn in gravel, or for other fishes which spawn in weeds beds.

ITQ 44 For trout, the regulation of numbers seems to depend on the number of territories available to fry when they start to feed; there is contest competition at this stage and after that the mortality seems to vary little from year to year for the rest of the fishes' lives. This fits in with some of Wynne-Edwards' ideas.

For perch and pike, the numbers in the adult year-classes seem to depend on the survival of very young fishes, and this seems to be related to weather factors, probably to water temperature. Presumably the weather factors affect the young fish through their food supply which is different from the food eaten by older members of the species. Once the young fish have survived this stage, it seems that their further expectation of life is limited only by the effect of predation; the perch in Windermere are probably regulated by the predation of pike on juvenile and young adult perch, and the pike population is now regulated by predation by the FBA on all pike that exceed about 60 cm in length.

ITQ 45 The number of eggs produced depends on the size of the fish, so the stunted populations will produce fewer eggs per individual than populations where the growth of individuals is not restricted. If the stunted population is very dense, the total production of eggs may be much the same as in a population which includes large individuals. The relationship between density and fecundity operates through the growth rate of the fish; this is limited by food supply.

ITQ 46 Breeding leads to increase in population. According to Wynne-Edwards, the birds regulate the number of eggs produced depending on the density of the population; but Lack believes that, on average, the birds lay the number of eggs that are likely to produce the largest number of juveniles. Lack considers that, for the majority of bird species, there is a density dependent mortality out of the breeding season (between the juvenile and breeding adult stages) and that this regulates the population numbers; Wynne-Edwards believes in contest competition at some stage during the life history resulting in ample food for the resident population. Lack believes that the mortality regulating population numbers in the majority of birds is starvation outside the breeding season. Wynne-Edwards would expect regulation between breeding adult population and eggs, and contest competition among juveniles and adults before breeding. Lack would expect regulation as a result of food limitation affecting juveniles and adults before breeding.

ITQ 47 Owls in Wytham Wood lay one, two or three eggs, or none; the actual number laid in any year seems to depend on the level of feeding of the female during the winter and up to the time of laying. Great tits usually lay eight to ten eggs, but can lay up to sixteen; the smallest clutch from undisturbed tits was five. The number of eggs depends on the nutritional level of the female tit, and this varies with a number of environmental factors, including weather. Red grouse lay up to ten eggs. At the population level, the mean number and quality of the eggs laid is related to the amount and quality of the heather in the autumn, winter and spring. The amount and quality of heather available to an individual hen depends on the size of territory taken by the cock and on subsequent changes in the heather due to winter weather, new growth in spring and grazing animals.

ITQ 48 For owls, failure of adults to breed was the key factor (and related to the availability of food for females in winter); failure of young owls to establish territories was a strongly density dependent factor that regulated the density of adults.

For great tits, 'mortality' outside the breeding season was the key factor; possibly starvation of juveniles accounts for much of this. Failure of adults to produce maximum numbers of hatched young (k_1 and k_2) was significantly density dependent, and this could be a regulating factor.

For red grouse, winter loss and losses at the egg and chick stages were both key factors at Kerloch, and both showed delayed density dependence. The quality of eggs laid by the females is related to food supplies during the previous autumn, winter and spring; territory size determines the population of breeding adults.

ITQ 49 Very low, since one chick is the maximum that one pair can produce in a year and the birds must be five years old before they start to breed. Actually, mortality for a group of Manx shearwaters breeding on Skokholm (South Wales) was probably less than 7 per cent for 1963 to 1964 and was 4 per cent for 1964 to 1965. This is a very low value indeed. The causes of death have not been explored in detail, but disease and predation (by gulls) seem probably not very important, so limitation of food supply could be a significant factor.

ITQ 50 The adult mortality was 30 per cent; the juvenile mortality was nearly 80 per cent. After the winter, the juveniles survived to the same extent as the adults.

ITQ 51 The animals that produce few young at a time are generally those which do not lie up in a burrow, lair or den, e.g. contrast the moose and the wolves. Most ungulates (mammals with hooves) produce a single young or twins; the new-born animal is able very soon after birth to run and to keep up with its mother and has well-developed senses—sight, hearing, smell. Animals with burrows or dens typically produce young that are helpless at birth, often with eyes closed, and hearing and smell probably poorly developed. The mother goes away to feed and returns to suckle the young which may grow rapidly and develop into alert individuals. There are exceptions: pigs, for instance produce many young at a birth.

ITQ 52 The moose has a reproductive life of twelve years and her potential is the production of twelve calves. The wolf has a reproductive life of eight years, but her potential is the production of forty-eight pups. If both animals fulfil their reproductive potential and their populations remain stable, the implication is that mortality among wolves is much higher than that among moose—one out of every twenty-four wolf pups must survive to breeding age (to replace male and female parents), whereas one out of every six moose calves must survive to breeding age. Thus forty-six of the wolf pups will die (spread over the eight years of reproductive life) and ten of the moose calves (spread over the twelve years of reproductive life).

ITQ 53 In the 60 weeks, the female vole could produce about nine litters of four young each, which equals thirty-six young. This is three times the number that the female moose is capable of producing in her much longer life, but the implication is that thirty-four of the vole's thirty-six young will die (spread over the fifty-six weeks of the vole's reproductive life).

ITQ 54 For voles, reproductive rates are higher in increasing populations than in peak and declining populations. Wildebeest in Ngorongoro produce calves more often than those on the Serengeti, suggesting that dense populations are less fecund than populations maintained at lower levels by predation. Moose on Isle Royale appear to have had a higher twinning rate in the 1960s, when the population was regulated by the wolves, than in 1929, when the population was much higher. Lions do not seem to show differences in birth rate, though the observed rates are lower than potential rates. Wolves have a low birth rate as a result of many females failing to breed and others producing fewer young than the maximum; the actual birth rate on Isle Royale was probably about one-quarter of the potential (giving about four wolf pups per female per generation time, the same figure as for the moose and voles).

ITQ 55 For voles, there is high mortality during the decline phase of populations both of juveniles and of breeding females, but so far the immediate causes of death have not been quantified. In phases of increase of population, there is considerable dispersal of individuals (which represents a loss to the resident population).

Wildebeest populations have a high mortality of calves as a result of predation, but much mortality may result from accidents such as calves becoming separated from their mothers. On Serengeti, many wildebeest probably die of malnutrition, in spite of the migrations of the herds; in Ngorongoro, hyena predation probably maintains the herds at a fairly constant level; whereas in Nairobi Park lion predation resulted in numbers falling from year to year (after a heavy mortality due to severe drought). For wildebeest, different mechanisms of population regulation operate in different areas.

For moose, there is a high mortality rate during the first year of life; much of this is due to predation by wolves, but a substantial proportion is due to accidents such as drowning. Mortality of older moose is largely the result of predation, but some animals reach advanced ages.

Among lions, there is high mortality during cub life due to abandonment, predation, violence or starvation. The population seems to be regulated by dispersal of juvenile lions, especially males, when they approach breeding age.

Wolf pups have a low rate of survival, probably dying mainly from starvation. Some dispersal of young adult wolves occurs, leading to the establishment of new packs.

ITQ 56 The total number of eggs produced at each spawning by fishes varies with their weight, being greater in larger fish: per kg of body weight, perch lay about 200 000 eggs, pike lay about 30 000, and trout about 1 600. These fish spawn annually in Britain. Contrast these intrinsic rates of increase with those of the moose and wildebeest, whose females generally produce one calf only each year (occasionally twins). Voles have high intrinsic rates of increase for mammals, but they are low compared with fishes.

ITQ 57 From 1952 to 1956, the rate of increase approximates to doubling of numbers every year; after 1956, the rate of increase is much less, falling to zero in about 1960 and then fluctuating between negative, positive and negative low rates.

ITQ 58 Insects that have been introduced as agents of biological control of noxious weeds include: the moth *Cactoblastis* which 'controlled' prickly pear *Opuntia* in Australia; the beetle *Chrysolina* which 'controlled' klamath weed *Hypericum perforatum* in USA (see *IPE*, chapter 9).

That such 'control' has been effective means that the herbivorous insect population regulates the numbers of individuals of its food plant population.

ITQ 59 Good examples are the carnivorous and 'parasitic' insects (insect parasitoids) which have served as successful agents of biological control, e.g. the ladybird beetle *Rodolia* on cottony cushion scale *Icerya* in California; the fly *Cyzenis* on winter moth *Operophtera brumata* in Canada (see *IPE*, chapter 9).

ITQ 60 Examples are: pike in Windermere feed predominantly on perch during the summer although other fish species are present in the lake; lions take wildebeest in Serengeti and in Nairobi Game Park, at-times when these are available, in preference to other ungulates (buffalo, zebra) also present.

ITQ 61 The moose on Isle Royale showed great fluctuations in population, with over-grazing, before the arrival of wolves. Other large herbivore populations, without natural predators, have demonstrated increases to levels where there were mass deaths (e.g. the deer on the Kaibab Plateau).

ITQ 62 The populations of red grouse in Scotland and of voles in Indiana appear to have regulatory mechanisms in which carnivores are not necessary.

ITQ 63 If the population of prey is reasonably stable, this allows exploitation by predators whose breeding strategy and/or general behaviour are such that their increase in numbers is adjusted to prey density. If the prey population undergoes large fluctuations, the predators may alternate between over-production and under-utilization, with heavy and low mortalities in subsequent years.

SAQ answers

SAQ 1 The curves are shown in Figure 22A and B. Compare them with Figures 3 and 4 of Unit 8 and read the comments on these two Figures.

SAQ 2 (a) gives a diagonal survivorship curve (3).
(b) is a horizontal and real cohort (4).
(c) is a positively skew rectangular curve (1).
(d) is a negatively skew rectangular curve (2).
(e) is one method of constructing a vertical imaginary cohort (5).

If you had difficulty with any of these, turn back to Section 8.1 and to the answers to ITQs 1 and 3.

SAQ 3
Andrewartha and Birch's theory includes as basic concepts: E, J.

Lack's theory includes as basic concepts: A, G.

Wynne-Edwards' theory includes as basic concepts: C, D, H, K.

The concepts B, F and L are not basic to any of the three theories.

If you had difficulty with any of these turn back to Section 9.0.

SAQ 4 From examination of Table 19 you will have noticed that the numbers of fish at all stages in year-class II is of the order of 100 times higher than the corresponding numbers in the three other year-classes; of these others, year-class III has higher numbers than IV, which has higher numbers than I.

(a) The calculation is shown in Table 21.

TABLE 21 Mortality factors calculated from data in Table 19.

	I	II	III	IV
log E	6.18	8.50	6.86	6.78
k_E for egg mortality	0.70	0.30	0.68	0.70
log F	5.48	8.20	6.18	6.08·
k_F for fry mortality	1.00	0.30	1.00	0.97
log S	4.48	7.90	5.18	5.11
k_S for mortality in the sea	1.48	1.60	1.48	1.51
log A	3.00	6.30	3.70	3.60
K for year class	3.18	2.20	3.16	3.18

For year-classes I, III and IV, the k-values for egg (and alevin) mortality is about 0.7, the k-value for fry mortality is about 1.0 and the k-value for mortality in the sea is about 1.5. For year-class II, the first two k-values are each only 0.3 (representing halving of the populations of eggs and of fry) but the third k-value at 1.6 is of the same order as the mortality in the sea of the other three year-classes.

The year-class mortality K is about 3.2 (reduction by more than one thousandfold) for year-classes I, III and IV but, for year-class II, K is 2.2, so the total mortality is proportionally ten times less than for the other year classes.

The large year-class II suffers relatively less mortality in freshwater stages than the other three ($k_E + k_F = 0.6$ for II compared with about 1.7 for the others). There must be various different causes for mortality of eggs, alevins and fry; it is likely that one cause is predation and this could be from long-lived predators such as other fish or fish-eating birds. The numbers of these predators could not adjust within the year to a very high number of salmon fry in a year following a small year-class the previous year.

If the predators are adjusted in numbers to the small year-classes IV and I, any increase in predators following year-class II will come up against a limit imposed by the small year-class III, and predator population numbers are likely, therefore, to return to the previous low level.

The phenomenon of year-classes with very high numbers occurring regularly with year-classes of much lower numbers between them is called 'cycle dominance' and is found in some populations of several different species of Pacific salmon. For sockeye, there is good evidence that predators (rainbow trout and birds) normally account for much of the freshwater mortality. Note that these salmonid fishes do not take up territories; the fry live in lakes and form large shoals feeding on zooplankton.

(b) The apparent fecundity can be found by dividing the number of eggs by the number of adults of that year-class (since the numbers repeat regularly every four years which is the time taken by an egg to become an adult). Multiply this number by 2 since half the fish are males. The values are: year-class I $15 \times 2 \times 10^2 = 3\,000$ eggs per female fish; year-class II 320 eggs; year-class III 2 920 eggs; year-class IV 3 000 eggs per female fish.

(c) The apparent fecundity of year-class II fish is about one-tenth of that of year-class I (320 compared with 3000). This suggests that the number of fish which succeed in spawning are only one-tenth of the population which would spawn if fishermen took the same proportion of year-class II fish as they do of the other year-classes. Thus, fishermen must take a considerably higher proportion of the adult salmon of year-class II than they take of the other year-classes. If fishermen take 90 per cent of year-class I, they probably take 99 per cent of year-class II. In fact, they know when a large year-class is due, and fish it with great enthusiasm.

Note that the regularity of the life cycle means that the four different year-classes of salmon living in the river are quite separate genetically and each can be treated as a separate 'population' with a four-year life cycle.

SAQ 5 (a) These are total population data. The figures for each year include many 'real' cohorts because adult buffaloes commonly live for ten years or longer; each year's figures can be thought of as a composite life table covering an 'imaginary' cohort. The decrease in numbers due to juvenile mortality represents deaths of members of a real cohort, those calves born that year.

(b) See Table 22 for the calculations. K increases (from 0.14 to 0.20) when the years are in correct order; k_1 is almost constant from year to year; k_2 is least in 1965 and highest in 1967 and 1970; k_3 is higher in 1969 and 1970.

(c) K for 1969 and 1970 must be at a level at which the population remains stable. The higher K for the two later years can be accounted for by increase in k_2 and k_3, i.e. higher mortality of juveniles and adults in the later years. Since k_2 is as high in 1967 as in 1970, it seems possible that k_3 is the mortality which regulates the population.

TABLE 22 Mortality factors calculated.

	1965	1967	1969	1970
log T	1.71	1.77	1.83	1.85
k_1 (failure to calve)	0.04	0.03	0.04	0.04
log C	1.67	1.74	1.79	1.81
k_2 (juvenile mortality)	0.06	0.10	0.08	0.10
log A	1.61	1.64	1.71	1.71
k_3 (adult mortality)	0.04	0.03	0.06	0.06
log S	1.57	1.61	1.65	1.65
K (total annual mortality)	0.14	0.16	0.18	0.20

(d) The success of the model in predicting the population changes suggests that k_3, the mortality of adult buffaloes, is density dependent and regulates the population.

(e) Presumably the population was determined in a different way before 1965. The real population did not increase as fast as predicted by the model, suggesting either a lower fertility or increased mortality other than k_3.

(f) It seems that k_2, the juvenile mortality, is not density dependent and is not a regulating factor. In fact, k_2 is the key factor. Since female buffaloes produce one calf only each year, the number on which k_2 really acts is less than half the number of adults, so fluctuations in total numbers from year to year will not be more than thirty per cent; in these years it was twenty per cent or less. The adult mortality, k_3, tends to restore the numbers towards the equilibrium level.

(g) Sinclair used the information to modify his first model by increasing the constant k_2 by a factor of 1.76; k_1 remained constant and k_3 was density dependent as in the earlier model. With this change in the average value for juvenile mortality, the predicted population levels for years before 1964 fitted the observed data very closely.

To sum up: the buffalo population level in Serengeti appears to be determined by juvenile mortality and regulated by adult mortality. The effect of rinderpest on the population was probably very similar to the effect of rinderpest on wildebeest—the level of juvenile mortality was almost doubled. When the rinderpest epidemic ended, the buffalo population increased towards a new equilibrium with a value of about 5.6 animals km^{-2}.

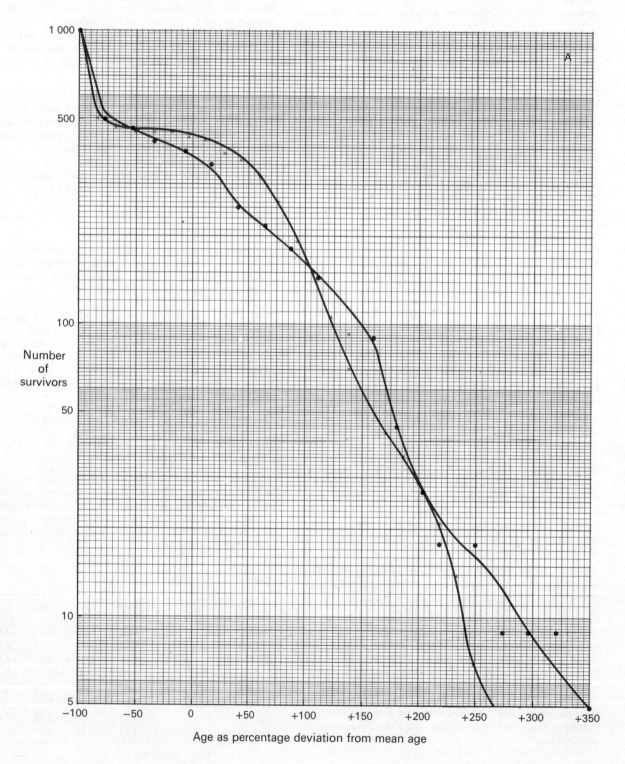

Figure 22 Survivorship curves for male (×) and female (●) buffaloes plotted against age expressed: A as percentage deviation from the mean length of life; B as percentage of the total life span.

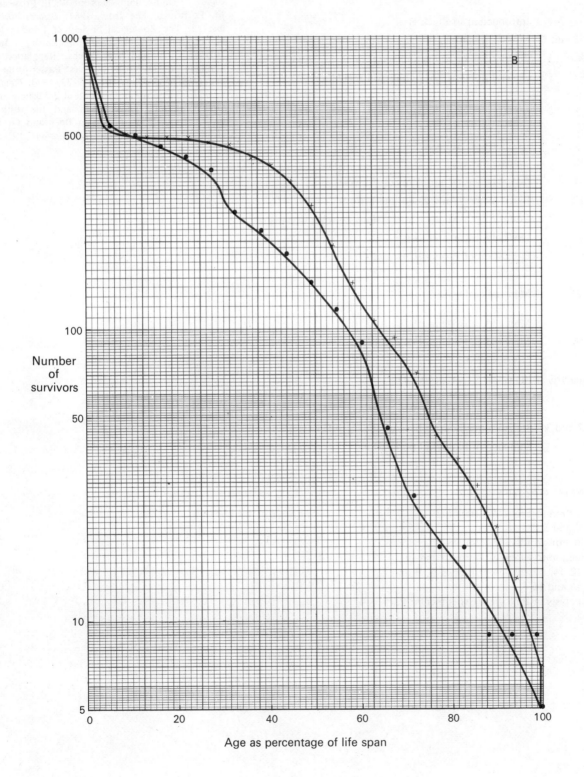

Age as percentage of life span

S323 ECOLOGY

Block A	**Energy flow through ecosystems**
	Introduction to Block A
Unit 1	Primary production—the process
Unit 2	Primary production in ecosystems
Unit 3	Consumers in ecosystems
Unit 4	The decomposers
Unit 5	Whole ecosystems
Block B	**Population dynamics**
	Introduction to Block B
Units 6 and 7	Insect populations
Unit 8	Populations of plants and vertebrates
Unit 9	General principles of population regulation
Block C	**Individuals and communities**
	Introduction to Block C
Unit 10	Climatic factors
Unit 11	Edaphic factors; problems of dispersal
Unit 12	Biotic factors: food chain interactions
Unit 13	Competitive and other interactions between organisms
Unit 14	Diversity in communities
	Discussion of Blocks A, B, and C
Block D	**Man and ecology**
Unit 15	Man and ecosystems
Unit 16	Man and communities of organisms

The Analysis of Biological Experiments (*ABE*)

Part 1	The rationale of biological experiments
Parts 2 and 3	Statistical methods in biological experiments
Part 4	Sampling of biological populations

Bindings are arranged as follows:

Introduction to Block A, Units 1–3
Units 4 and 5
Introduction to Block B, Units 6–9
Introduction to Block C, Units 10 and 11
Units 12–14, Discussion of Blocks A, B and C
Units 15 and 16
The Analysis of Biological Experiments: Parts 1–4

Acknowledgements

Grateful acknowledgement is made to the following sources for material used in these Units:

Figure 3: The Stony Brook Foundation and the author for E. S. Deevey jr. in *Quarterly Review of Biology,* **22,** 283–314, 1947; *Figures 5 and 6 and material in Section 8.2.1:* Dr. J. Sarukhan for his PhD thesis, University of Wales, 1971; *Figures 7 and 9:* E. Schweizerbart'sche Verlagsbuchhandlung for E. D. Le Cren in *Verh. int. Verein Limnol,* **12,** 187, 1955 and E. D. Le Cren in *Mitteilungen Internationale Vereinigung fur Theoretische und Angewandte Limnologie,* **13,** 88, 1965; *Figure 8:* Collins Publishers for W. E. Frost and M. E. Brown, *The Trout,* 1967; *Figures 10, 11 and 12:* The Zoological Society of London for H. N. Southern in *Journal of Zoology,* **162,** 197–285, 1970, and J. R. Krebs, 'Regulation of numbers in the great tit (Aves: Passeriformes)' in *Journal of Zoology,* **162,** 317–33, 1970; *Figures 13 and 14:* Centre for Agricultural Publishing and Documentation, Wageningen, and the author for C. J. Krebs in P. J. den Boer and G. R. Gradwell (ed.), *Dynamics of Populations,* 1971.